# Form of Contract

## Lump sum contracts

The Red Book

Fifth edition
2013

**Published by**
**Institution of Chemical Engineers (IChemE)**
**Davis Building**
**165–189 Railway Terrace**
**Rugby**
**Warwickshire**
**CV21 3HQ, UK**
IChemE is a Registered Charity in England and Wales and a charity registered in Scotland (SCO39661)



ISBN: 978-0-85295-542-0

Fifth edition 2013
Reprinted 2014

## Agreement on Website

To facilitate use of the Form of Contract the text of the Agreement is freely available on the IChemE website at www.icheme.org

Typeset by Techset Composition Ltd, Salisbury, UK

Printed by Bell & Bain, Glasgow, UK

# Contents

## Appendices



## Index

# Contracts Committee members

Gordon Bateman (Chairman)
Paul Buckingham
Roger Button
John Challenger
Geoffrey Hawker
Chris Hillier
David McLurgh
Henry Rowson
Jon Shattock
Ray Shaw
Iain Turner
Jim Williams

Secretary to the Contracts Committee: Tracey Donaldson, IChemE

# Introductory notes

## 1. Use of the Conditions

These conditions, referred to as the Institution of Chemical Engineers (IChemE) 'Red Book', are specifically published for lump sum contracts for the design, construction and commissioning of performance based process plants.

The Red Book contains a typical form of Agreement and a set of General Conditions of Contract for lump sum contracts. Guidance is provided on how to compile the Specification and the Schedules to which reference is made in the General Conditions. Guide Notes A to S are intended to aid interpretation of some of the General Conditions and to explain the need for certain Special Conditions. Users of the Conditions are strongly advised to read the relevant Guide Notes before preparing the Specification, the Schedules and any Special Conditions, all of which will have to be written individually to reflect specific contract requirements. Some of the Schedules may be optional.

The following Forms of Contract are also published by IChemE:

- Green Book (cost reimbursable)
- Burgundy Book (target cost)
- Yellow Book (subcontract)
- Brown Book (civil subcontract)
- Orange Book (minor works).

## 2. General

The Conditions have been developed by the IChemE Contracts Committee, the members of which represent a broad spectrum of those involved in design, commercial, legal, construction, ownership and operation of process plants both in the UK and in many other parts of the world. The General Conditions in particular have been formulated to reflect best practice and relationships within the process plant sector, which are generally far less adversarial in nature than in other parts of the construction industry. IChemE has always adopted the basic philosophy that the parties should co-operate to achieve the mutual objective of a successful project, rather than regarding the Contract as simply the basis for an adversarial relationship. IChemE believes that it is in the best interests of the parties to deal fairly with each other and with their Subcontractors, specialists and suppliers in an atmosphere of co-operation in order to achieve 'win-win' solutions to the problems that will inevitably arise during the course of a project.

Anyone contemplating modifications to the General Conditions should take care not to introduce provisions which may conflict with these well-established practices and relationships. They should particularly be aware of the risk of introducing inconsistencies within the Conditions, or provisions that may be unenforceable.

Users are cautioned that IChemE has not authorised or approved any amendment unless it is published by IChemE.

The Conditions have been prepared with process plants and manufacturing facilities in mind. While they are certainly suitable for other types of projects and are increasingly being adopted in a range of performance related projects for such diverse industries as nuclear, pharmaceuticals, water, tunnelling and high voltage distribution, such applications may require the preparation of Special Conditions to reflect the legislation and particular practices that apply to those sectors.

The General Conditions incorporate at Clause 41 payment provisions which are consistent with the requirements of the Housing Grants, Construction and Regeneration Act 1996, Part II, as amended by Part 8 of the Local Democracy, Economic Development and Construction Act 2009. Clause 47 provides for adjudication in accordance with those Acts.

The Conditions are concerned with a project carried out within the United Kingdom and subject to English law and this is reflected in the drafting of this Contract

A full range of International Contracts, using the same approach and general format as the UK editions but suitable for a wide range of jurisdictions, is published by IChemE.

## 3. Choice of contract–lump sum, cost reimbursable or target cost?

Contracts based on 'lump sum' and wholly 'cost reimbursable' payment are the extremes of a wide range of payment mechanisms that can be used to cover the distribution of risks and responsibilities between the parties to an engineering and construction contract for the realisation of a process plant. Whilst at first sight it would appear that the Purchaser's best interests would be served by the use of a contract based upon a lump sum price for his project, he should satisfy himself that he has properly examined the alternatives and assessed their relative advantages and disadvantages.

The essence of a lump sum contract is that, in return for an agreed firm price, the Contractor provides all that is necessary to ensure that the finished plant complies with the Specification, achieves the required levels of performance and is completed on time. The Specification, performance requirements and guarantees, and all other technical and commercial contractual conditions must be agreed in detail before the Contract is awarded. Once the Contract becomes effective, the lump sum price is payable regardless of the actual costs incurred by the Contractor. The Contractor must, therefore, include contingencies in his price for estimating errors, uncertainties such as cost escalation and currency fluctuations, costs which on the basis of his experience will inevitably be incurred from time to time. Having included for these contingencies, the eventual cost of which may turn out to be significantly different from his estimates, the Contractor takes the risk that over-expenditure will result in a smaller profit than expected or even a loss, in the knowledge that, if all goes well, his profit may be greater than expected. Conversely, whilst the Purchaser does not carry the risk of cost overruns, he cannot benefit from any savings.

In contrast, the essence of a reimbursable contract is that in providing the goods and services that the Purchaser requires the Contractor is reimbursed for all the costs he incurs including, subject to the Contract and provided he exercises normal professional skill, corrective costs arising from the inevitable design and construction errors. The Purchaser gains the benefits of any engineering, material or construction cost savings but bears the risk of any cost over-run.

There is a range of payment mechanisms between the two extremes which may form the basis of a contract. These include:

- reimbursable cost plus a percentage fee;
- reimbursable cost plus a fixed fee;
- target cost (shared over-run and/or under-run)
- unit rate (including remeasure);
- lump sum services and materials with reimbursable construction;
- guaranteed maximum price;
- lump sum.

Only the first two of these are essentially reimbursable but all except lump sum involve, to some extent at least, reimbursement to the Contractor of the costs he actually incurs.

A target cost contract is a particular type of cost reimbursable contract in which the contractor is reimbursed his costs subject to the application of a formula which allows the contractor to share in savings made, often called 'gain share', or to contribute towards additional costs incurred, called 'pain share', according to how well the parties are able to manage the cost of the works. The risk allocation of a target cost contract is such as to place more risk on the Contractor than a pure cost reimbursable contract, but not as much risk as a price-based contract. The particular risk allocation will depend upon the approach taken in the target cost formulae and associated payment provisions. Thus in terms of risk allocation a target cost contract lies somewhere between a lump sum contract and a cost reimbursable contract.

Table 1 compares the advantages and disadvantages of lump sum, target cost and reimbursable contracts and provides guidance upon which the choice of contract form may be based. If the Purchaser elects to place a lump sum contract it is essential that he provides potential contractors with adequate information upon which to base their estimates and sufficient time to submit their tenders.

## 4. Pre-contract activities

Lump sum prices in bids can only be properly compared when each covers the same scope of supply and preferably in a similar format, usually as defined in the invitation to bid documents. If each tenderer submits its proposal in its own standard format, the task of making a comparison can be difficult, prolonged and prone to error.

Comparison of tenders for lump sum contracts will require all important criteria to be fully defined and will therefore need considerable resources of experienced personnel and time to compile properly and accurately.

The tender invitation should therefore incorporate a draft contract document including, as a minimum, drafts of the Specification and schedules containing the Purchaser's information and requirements:

- description of the Works;
- responsibilities of Purchaser;
- applicable standards;
- health, safety and environment;
- quality assurance and validation (if applicable);
- subcontracting;
- training by Contractor;
- times of completion;
- liquidated damages for delay;
- performance guarantees and damages for failure;
- terms of payment.

Outlines of the other Schedules should also be included to be completed by the tenderers.

The better the definition in the tender invitation, the shorter will be the time needed for clarification and evaluation of the tenders. For example, if the Purchaser wishes to benefit from the application of Schedule 10 (Parts with limited working life and spare parts) this should be made clear in the tender invitation.

A similar consideration applies to factors which will affect the risks of the Contractor who will eventually be appointed. These include the amount to be included in paragraph 4 of the Agreement and the rates and limits for liquidated damages (Schedules 12 and 17) which should be stated in the tender invitation, as should any proposal by the Purchaser to make incentive payments (see Guide Note K).

If the Purchaser would prefer to carry the risk of inflation, the method of cost adjustment he proposes should be indicated in the invitation to tender in broad terms only, the tenderer being invited to submit his own proposals.

If the tenderer cannot accept the Purchaser's requirements in full or if there is any doubt about the compatibility of his own proposals and those of the enquiry, he should bring this to the Purchaser's attention at or before the time of tendering. The points at issue should be discussed and a course of action agreed. Tender evaluation is assisted if in such cases the tenderer finds it possible to submit alternatives incorporating his own proposals in addition to a tender based exactly on the terms of the enquiry.

The Purchaser may, if the circumstances justify, elect to have each tenderer provide a bid bond in order to ensure that the successful tenderer enters into a contract or pays the Purchaser his expenses of abortive negotiations. The bond may be payable on a written demand stating that the tenderer has been awarded the Contract but within a stated number of days has failed to enter into a binding contract (see Guide Note Q).

## 5. Contract procedure

It is strongly recommended that the Agreement be used in the form provided. The Agreement defines exactly which documents make up the Contract. Documents dated earlier than the date of the Agreement have no effect unless specifically incorporated in the Contract nor would any other terms and conditions, for example those sometimes printed on the back of order forms or quotations.

If any other form of agreement is to be used the Purchaser must ensure that all the statements shown in the Agreement are included in the contract documentation in some other way. It is essential that all the necessary supporting documents, including the General and Special Conditions of Contract, with the accompanying Schedules, Specification and any other necessary contractual and technical documents are incorporated. There are frequent references to the Agreement in the General Conditions.

When there have been any changes in concept during pre-contract negotiations, any relevant Special Conditions or Schedules should be amended before the Agreement is signed. This will include incorporating in the Special Conditions, the Specification or the Schedules all those documents of

**Table 1**—Comparison of lump sum, target cost and reimbursable contracts

| LUMP SUM | TARGET COST | REIMBURSABLE |
|---|---|---|
| **Project definition required in enquiry** | | |
| • Complete in terms of overall Purchaser's requirements together with all relevant supporting information and data. | • Where the Target Cost is included in the contract, the same requirements as for lump sum.<br>• Where the Initial Target Cost is to be agreed post-contract, as much information as is available/possible. | • Sufficient to enable tenderers to demonstrate that they have the skills and resources needed. |
| **Advantages** | | |
| • Purchaser knows his expenditure commitment.<br>• Allows fullest competition between potential contractors.<br>• Cost comparison between bids is relatively straightforward. | • Possibility of 'gain share' for Purchaser and Contractor if costs of works are well managed and delivered within the Target Cost, but conversely 'pain share' if there are problems.<br>• Greater sharing of risk between the parties in comparison to lump sum option.<br>• Lower level of cost associated with risk could be built into the Contractor's tender.<br>• Potentially shorter bidding time and lower bidding costs than for lump sum. | • Requires limited enquiry definition.<br>• Shortest bid time.<br>• Complete flexibility—design development and Purchaser participation practicable.<br>• Purchaser/Contractor conflict of interest is minimised.<br>• Purchaser has more control over the work.<br>• Purchaser can assess tenderers' rates.<br>• Purchaser can use Contractor to evaluate alternative schemes and carry out a range of studies.<br>• Purchaser can terminate at will without incurring substantial costs.<br>• Competition can continue for subcontract packages (when developed). |
| **Disadvantages** | | |
| • Purchaser/Contractor interests are more divergent than in other forms of contract.<br>• Lengthy enquiry preparation time to prepare clear and complete. specification and bid documentation.<br>• Long bidding time is required (2–4 months or more).<br>• Crucial process design phase compressed into a short period.<br>• Lack of flexibility—changes are difficult/expensive.<br>• Purchaser participation in project is difficult and limited.<br>• Cost to Purchaser may be increased due to contingencies for risk and escalation.<br>• Emphasis on low bid price in bid evaluation may divert attention from technical merit and experience (ie failure to evaluate value offered). | • More risk on the Contractor than a pure cost reimbursable contract.<br>• Significant effort required and cost incurred to ensure that the administrative requirements of the Contract are managed appropriately.<br>• Bid evaluation may be difficult as it will generally be based on Contractor's relevant record on similar projects, and quality of available resources and rates. | • Contractor has no monetary incentive to minimise cost to Purchaser.<br>• Purchaser has no assurance of final cost.<br>• Purchaser has to check and verify Contractor's man-hour and expense records.<br>• Bid evaluation may be difficult as it will generally be based on Contractor's relevant record on similar projects, and quality of available resources and rates. |
| **Typical applications** | | |
| • When competing tenderers offer different proprietary processes.<br>• When Purchaser provides tenderers with front-end engineering packages and detailed specifications.<br>• 'Standard package' plants. | • When the initial specification/scope cannot be fully defined and/or supported by relevant project specific data and information. | • Where flexibility is required, such as when project cannot be defined in detail at the start because not yet fixed. |

**Table 1**—(*Continued*)

| | | |
|---|---|---|
| • When dictated by financing requirements. | • When requirement offers the potential for technical innovation (design, engineering and/or manufacture) during the term of the Contract but a fully reimbursable approach is not acceptable or when it is appropriate to offer a 'reward' for innovation.<br>• When the flexibility of the reimbursable contract is to be combined with the discipline of a Target Cost. | • When Purchaser wishes to participate extensively in the design—for example, a development project, especially one dependent on Purchaser know-how.<br>• When confidentiality considerations preclude the issue of detailed enquiry specifications to tenderers.<br>• When Purchaser considers contingencies for risk in a lump sum are likely to be excessive.<br>• When potential contractors are not prepared to take lump sum contracts (for example, very large contracts, times of high inflation or currency fluctuation, times of political uncertainty). |
| **Comments** | | |
| • Complete project definition is essential for a lump-sum contract.<br>• Tenderers' tendering costs are very high. Purchaser should minimise contracting industry tendering costs (and thus plant costs) by<br>(a) not inviting bids until there is a high probability of the project proceeding,<br>(b) minimising the number of tenderers,<br>(c) pre-qualifying each tenderer so that he will not be rejected later on grounds that should have been known to Purchaser before the enquiry was issued, and<br>(d) considering reimbursing pre-qualified unsuccessful tenderers for their costs of bid preparation. | • Whilst initial complete project definition is not always essential, the risk associated with any shortfalls needs to be quantified and assessed such that it can be properly considered and allocated.<br>• The Purchaser must have experienced, competent in-house project management capability to interface adequately with the Contractor. | • The most flexible type of contract, allowing a very rapid start.<br>• Flexibility may encourage Purchaser to introduce design changes, resulting in increased cost and longer programme.<br>• Contractor's profit/loss is limited.<br>• The Purchaser must have experienced, competent in-house project management capability to interface adequately with the Contractor. |

the Contractor's tender which it is agreed shall be part of the Contract.

## 6. Modifications to the Contract

Any part of a contract can be modified by agreement between the parties at any time after it has been signed, but any modification must be expressed in unambiguous terms in a document signed by both parties.

The Conditions are intended for projects which are fully defined at the time that the Contract is placed. However, instances often arise when the bulk of the project may be defined but some elements are still open to development or change for reasons outside the Contractor's responsibility or knowledge. In instances such as these it is feasible to adopt the Conditions provided these elements are clearly defined in Special Conditions and, when appropriate, in an additional Schedule. It will also be necessary to state the applicable payment mechanism for any work not included within the lump sum contract price.

# Agreement

*(See Guidance on completing the Agreement)*

**AN AGREEMENT** dated *[day, month and year in words]*

between *[full name]* ('the **Purchaser**') of *[address]*

and *[full name]* ('the **Contractor**') of *[address]*

**WHEREAS**

The **Purchaser** wishes to have a process plant to be known as *[name of Plant ]*

constructed at *[location of Site]*

and wishes the **Contractor** to carry out and complete the **Works** defined in the **Contract** and the **Contractor** is willing and able to carry out and complete the **Works** in accordance with the **Contract**.

**THIS AGREEMENT** provides as follows:

1. The following documents and their attachments (if any) shall together form the contract between the **Purchaser** and the **Contractor** and the term **'Contract'** shall in all such documents be construed accordingly:

   (a) This **Agreement**.

   (b) The **Contract Conditions**, comprising:

   Part I – The **General Conditions**, being Clauses 1–49 as set out in the IChemE Form of Contract for Lump Sum Contracts (the 'Red Book'), 5th edition, 2013,

   and

   Part II – The **Special Conditions** (if any)

   (c) The **Specification**.

   (d) The **Schedules**:

   **Schedule 1**: Description of the Works;
   **Schedule 2**: Documentation;
   **Schedule 3**: Responsibilities of Purchaser;
   **Schedule 4**: Health & Safety;
   **Schedule 5**: Environmental protection & waste disposal;
   **Schedule 6**: Quality assurance and validation;
   **Schedule 7**: Subcontracting;
   **Schedule 8**: Contractor's named personnel;
   **Schedule 9**: Training by Contractor;
   **Schedule 10**: Parts with limited working life and spare parts;
   **Schedule 11**: Times of completion;
   **Schedule 12**: Liquidated damages for delay;
   **Schedule 13**: Pre-installation tests and procedures;
   **Schedule 14**: Criteria for the completion of construction;
   **Schedule 15**: Take Over procedures;
   **Schedule 16**: Performance tests and procedures;
   **Schedule 17**: Performance guarantees and damages for failure;
   **Schedule 18**: Valuation of Variations and claims;
   **Schedule 19**: Contract Price and payment.
   **Schedule 20**: Contract co-ordination;
   **Schedule 21**: Reports and records.

   For the purpose of identification, the contents of the **Contract**, including the number of pages in each part, are listed in the Annex to this **Agreement** attached hereto.

2. The **Contract** constitutes the entire agreement between the **Purchaser** and the **Contractor** with respect to the performance of the **Works** and supersedes any prior negotiation, representation or agreement relating thereto, whether written or oral,

except to the extent that they are expressly incorporated in the **Contract**. No change, alteration or modification to the **Contract** shall be effective unless the same shall be in writing and signed by both parties.

3. The **Contract Price** is the sum of *[name(s) of currency(ies) and amount(s) in figures and words].*

4. The total aggregate liability of the **Contractor** to the **Purchaser** as provided in Sub-clause 45.3 shall not exceed *[name(s) of currency(ies) and amount(s) in figures and words].*

5. The **Contractor's** liability in respect of:

   (a) Loss of or damage to property of the **Purchaser** and his **Affiliates** in accordance with Sub-clause 30.13 of the **General Conditions** shall not exceed *[name(s) of currency(ies) and amount(s) in figures and words].*

   (b) The amount to be stated in accordance with Sub-clause 37.12 of the **General Conditions** is *[name(s) of currency(ies) and amount(s) in figures and words].*

6. The **Deductible** shall be *[name of currency and amount in figures and words].*

7. In case of conflict between any of the documents accompanying this **Agreement**, the order of precedence shall be as stated in Sub-clause 1.2 of the **General Conditions**.

8. For the purposes of Sub-clauses 6.3, 7.3, 8.3 and 28.3 of the **General Conditions**, the date of the **Contractor's** tender shall be *[date].*

9. The date for the commencement of the **Works** shall be *[date].*

10. The **Purchaser** hereby appoints *[name]* to act as the **Project Manager** for the purposes of the **Contract**.

11. The **Contractor** hereby appoints *[name]* to act as the **Contract Manager** for the purposes of the **Contract**.

12. The bank whose base lending rate is referred to in Sub-clause 1.1 of the **General Conditions** in the definition of **Agreed Rate** is *[name of bank or where more than one currency is to be used, the names of the banks and a statement as to which base lending rate applies to which currency].*

13. Wherever **Profit** is expressly referred to in the **General Conditions** it shall be ....% of the applicable **Cost**.

14. Any **Notice** to be served in accordance with Sub-clause 1.8 of the **General Conditions** shall be sent to the postal address or fax number stated below:

| | | |
|---|---|---|
| The **Purchaser**: | For the attention of: | *[name and/or position]* |
| | Address: | *[full postal address]* |
| | Fax Number: | *[fax number]* |
| The **Contractor**: | For the attention of: | *[name and/or position]* |
| | Address: | *[full postal address]* |
| | Fax Number: | *[fax number]* |

15. The **Contract** shall be governed by the laws of England and Wales.

16. The seat of any arbitration proceedings under Clause 49 shall be *[place].*

**SIGNED** by the duly appointed representatives of the parties:

| | |
|---|---|
| For and on behalf of *[full name of the **Purchaser**]* | For and on behalf of *[full name of the **Contractor**]* |
| Signature: | Signature: |
| Name: | Name: |
| Position: | Position: |

# General Conditions of Contract

## 1. Definitions, interpretation, waiver and Notices

1.1 In the **Contract**, unless the context otherwise requires, the following expressions shall have the following meanings.

**'Acceptance Certificate'** has the meaning stated in Sub-clause 36.3 and **'Accept'**, **'Acceptance'** and **'Accepted'** shall be construed accordingly.

**'Affiliate'** means any entity that directly or indirectly controls, is controlled by, or is under common control with another entity. For the purposes of this definition, control means having the beneficial ownership of more than 50% of the issued share capital of a company or the legal power to direct or cause the direction of the general management of the company, and 'controls' and 'controlled by' shall be construed accordingly.

**'Agreed Rate'** means the base lending rate(s) as may be current from time to time of the bank(s) stated in the **Agreement**.

**'Agreement'** means the document titled 'Agreement' signed by the **Purchaser** and the **Contractor** and which refers to these **General Conditions**.

**'Applications Software'** means software that has been developed specifically as part of the **Works** including configuration of **Third Party Software** or **Contractor's Software**.

**'Approved Programme'** means the programme of work approved from time to time in accordance with Clause 13 (Times of completion and Approved Programme).

**'Confidential Information'** has the meaning stated in Sub-clause 20.1.

**'Construction Completion Certificate'** has the meaning stated in Sub-clause 32.2.

**'Contract'** has the meaning stated in the **Agreement**.

**'Contract Conditions'** means the **General Conditions** and the **Special Conditions**.

**'Contract Manager'** means the individual named as such in the **Agreement** subject to Clause 12 (Contract Manager and Contractor's staff).

**'Contract Price'** means the sum stated in the **Agreement**.

**'Contractor'** means the person named as such in the **Agreement** or his permitted assigns.

**'Contractor's Equipment'** means all equipment, construction plant, vehicles, temporary buildings and offices, materials, tools or things brought on to the **Site** by the **Contractor** or a **Subcontractor** for carrying out the **Works** but not for permanent incorporation in the **Plant**, excluding **Temporary Works**.

**'Contractor's Software'** means software owned by the **Contractor**.

**'Cost'** means any cost properly incurred by the **Contractor** in carrying out any of his obligations under the **Contract**, and **'Costs'** shall be construed accordingly.

**'Decision'** means any decision, certificate, notification, instruction, order, agreement, approval, rejection or consent.

**'Deductible'** means the sum stated in the **Agreement** being the applicable insurance deductible under the policy or policies of insurance provided in accordance with Sub-clause 31.1.

**'Default Certificate'** has the meaning stated in Sub-clause 44.6.

**'Defect'** means any work done or any **Materials** or the **Plant** or any part of it which does not comply with the **Contract**, provided that such matter shall not be a **Defect** if it is caused by:

(a) normal wear and tear;
(b) a failure by the **Purchaser** to operate and maintain the **Plant** in accordance with any operating and maintenance manuals provided by the **Contractor** and/or with good engineering practice; and/or
(c) a failure by the **Purchaser** to comply with any of his obligations under the **Contract**.

**'Defects Liability Period'** means a period of three hundred and sixty five days, or such other period as may be established in accordance with Sub-clause 37.9 or 37.11, after the date of the relevant **Take Over Certificate** or other commencement date as described in Sub-clause 37.5.

**'Documentation'** means any relevant documents in paper or electronic form, including drawings, technical software, images, designs, manuals or records.

**'Expert'** means a person referred to and so called in Clause 48 (Reference to an Expert).

**'Final Certificate'** has the meaning stated in Sub-clause 38.1.

**'Final Date for Payment'** has the meaning stated in Sub-clause 41.5.

**'Force Majeure'** has the meaning stated in Sub-clause 14.6.

**'General Conditions'** means these General Conditions of Contract.

**'IChemE'** means the Institution of Chemical Engineers, 165–189 Railway Terrace, Rugby, Warwickshire, UK, CV21 3HQ.

**'Legislation'** means all applicable laws, statutes, bye-laws, regulations and other measures having the force of law.

**'Materials'** means machinery, plant and other items of equipment and materials intended to form part of the **Plant** and other things needed in its operation excluding software, to be supplied by the **Contractor** under the **Contract**.

**'Notice'** means a formal notice issued by the **Purchaser** to the **Contractor** or by the **Contractor** to the **Purchaser** in accordance with Sub-clause 1.8.

**'Notice of Adjudication'** means a **Notice** as described in Sub-clause 47.2.

**'Notice of Arbitration'** means a **Notice** as described in Sub-clause 49.1.

**'Notice of Dispute'** means a **Notice** as described in Sub-clause 46.4.

**'Notice of Termination'** means a **Notice** as described in Sub-clause 43.1.

**'Performance Test Commencement Period'** means the period within which the performance tests are to commence as stated in **Schedule 11** (Times of completion) or any extension thereof granted in accordance with Clause 35 (Performance tests).

**'Performance Test Period'** means the period within which the performance tests are to be carried out and completed, as stated in **Schedule 11** or any extension thereof granted in accordance with Clause 35.

**'Plant'** means the plant as described in the **Specification** to be constructed at the **Site** including any software.

**'Profit'** shall be as defined in the **Agreement**.

**'Project Manager'** means the individual named as such in the **Agreement** subject to Clause 11 (The Project Manager).

**'Project Manager's Representative'** has the meaning stated in Sub-clause 11.5.

**'Purchaser'** means the person named as such in the **Agreement** or his permitted assigns.

**'Purchaser's Risk'** means:

(a) the use or occupation of the **Plant** (or any part) by the **Purchaser**, his personnel, consultants, or agents, or other contractors (not being employed by the **Contractor**); or

(b) any design or information provided by the **Purchaser** (other than any design or information which the **Contractor** is required to verify in accordance with his obligations under the **Contract**); or

(c) any wrongful or negligent act or omission of the **Purchaser**, his personnel, consultants, or agents, or other contractors (not being employed by the **Contractor**); or

(d) riot, war, invasion or hostilities (whether or not war be declared), terrorism, civil unrest, civil war, rebellion, revolution, insurrection or military or usurped power or similar events; or

(e) ionising radiations or contamination by radioactivity from any nuclear fuel or from any nuclear waste, from the combustion of nuclear fuel, radioactive, toxic, explosive or other hazardous properties of any explosive nuclear assembly or nuclear component thereof; or

(f) pressure waves caused by aircraft or other aerial devices travelling at sonic or supersonic speed.

**'Schedules'** means the schedules listed in the **Agreement** and forming part of the **Contract**.

**'Section'** means a part of the **Plant** identified in the **Contract** as a Section (if any).

**'Site'** means the area within which the **Plant** is to be constructed, together with all other areas as the **Contractor** shall be permitted to use in connection with the **Works**, as specified in **Schedule 1** (Description of the Works).

**'Site Manager'** has the meaning stated in Sub-clause 12.2.

**'Site Materials'** means (a) any **Materials** from the time of their delivery to the **Site**; and (b) any materials which the **Purchaser** free-issues to the **Contractor** for incorporation into the **Plant** from the time of their delivery to the **Site**.

**'Special Conditions'** means the conditions of contract entitled Special Conditions (if any).

**'Specification'** means the specification forming part of the **Contract**, which sets out the technical definition of the **Plant**.

**'Subcontractor'** means any subcontractor or supplier of any tier to whom the preparation of any design, the supply of any **Materials** or the carrying out of any other part of the **Works** is subcontracted.

**'Take Over Certificate'** has the meaning stated in Sub-clause 33.7, and **'Take Over'**, **'Taking Over'** and **'Taken Over'** shall be construed accordingly.

**'Temporary Works'** means all temporary works and structures of every kind constructed at the **Site** by the **Contractor** to facilitate construction of the **Plant**.

**'Termination Certificate'** has the meaning stated in Sub-clause 43.5.

**'Third Party Software'** means software owned by a third party.

**'Variation'** has the meaning stated in Sub-clause 16.1.

**'Variation Order'** means an order by which a **Variation** is ordered, or other notification made to the **Contractor** in accordance with the **Contract**.

**'Works'** means the design, engineering and other services to be provided by the **Contractor** including the provision and construction of the **Plant** and any **Temporary Works** and the subsequent dismantling or removal of the **Temporary Works** when no longer required, and any other work to be carried out by the **Contractor** in accordance with the **Contract**.

**1.2** The **Contract** documents shall be construed as mutually explanatory of one another. In the event of conflict between any of the documents comprising the **Contract**, the order of precedence shall be as follows:

(a) the **Agreement**;
(b) the **Special Conditions** (if any);
(c) the **General Conditions**;
(d) the **Specification**;
(e) the **Schedules**.

**1.3** The singular shall include the plural and the plural the singular except where the context otherwise requires and the words 'he', 'him' and 'his' shall be taken to mean 'she', 'her' and 'hers' where appropriate.

**1.4** No **Decision** required under the **Contract** shall be unreasonably refused or delayed.

**1.5** References to:

(a) 'day' shall mean a calendar day; and
(b) 'includes' or 'including' shall mean includes or including without limitation.

**1.6** Any communication in accordance with or in connection with the **Contract** shall be in writing, which shall include handwritten, typewritten or printed documents sent by hand, post, fax, email or other means resulting in a permanent record.

**1.7** The failure of either party to enforce any provision of the **Contract** shall not constitute a waiver of his right to require complete performance by the other party. A party shall only have waived his rights under the **Contract** if such waiver shall have been recorded in writing and signed by a duly authorised representative of that party. No waiver of any breach of the **Contract** shall constitute a waiver of any subsequent breach of the **Contract**.

**1.8** Any **Notice** to be served by one party on the other shall:

(a) be contained in a separate document and shall state the number(s) of the Clause(s) or Sub-clause(s) in the **Contract** under which the **Notice** is issued; and
(b) be sent to the person at the postal address or fax number stated in the **Agreement**, and shall be sufficiently delivered if marked for the attention of that person and:

  (i) left at the postal address stated in the **Agreement**; or
  (ii) sent to that address by Recorded Delivery or Special Delivery post; or
  (iii) sent by fax to the fax number stated in the **Agreement**.

In the case of a **Notice** left at the address stated in the **Agreement** such **Notice** shall be deemed to have been received on the day left at that address if left at or before 5.00 pm on a day on which banks in England were open for business, or if left after 5.00 pm such **Notice** shall be deemed to have been received on the next day on which banks in England were open for business.

In the case of a **Notice** sent by Recorded Delivery or Special Delivery post to the address stated in the **Agreement** or faxed to the fax number stated in the **Agreement**, such **Notice** shall be deemed to have been received on the day on which it was recorded as delivered by Recorded Delivery or Special Delivery post or by fax if delivered at or before 5.00 pm on a day on which banks in England were open for business, or if recorded as delivered after 5.00 pm such **Notice** shall be deemed to have been received on the next day on which banks in England were open for business.

Either party may at any time by **Notice** in writing to the other change the name of the person to whom a **Notice** should be addressed, the postal address or fax number stated in the **Agreement**.

**1.9** The headings and marginal notes in the **Contract Conditions** together with the accompanying Introductory Notes, Guidance on compiling the Agreement, its Annex, the Specification and Schedules, and Guide Notes shall not form part of the **Contract** nor shall they be taken into consideration in the interpretation of the **Contract**.

## 2. Co-operation

**2.1** The parties and the **Project Manager** shall each co-operate with each other in the discharge of their respective obligations under the **Contract** with the aim of satisfactorily completing the **Plant** and the **Works** in accordance with the **Contract**.

**2.2** The parties shall deal fairly, openly and in good faith with each other. Each party shall disclose information which the other might reasonably need in order to exercise his rights and to perform his obligations under the **Contract**. In particular, each party shall promptly disclose full information to the other concerning any matter which will or may prevent the **Plant** and **Works** being completed in accordance with the **Contract**. The parties shall work together in a manner consistent with their respective obligations under the **Contract** to resolve or mitigate the effect of any such matter.

## 3. Contractor's responsibilities

**3.1** In consideration of payment by the **Purchaser**, the **Contractor** shall regularly and diligently carry out and complete the **Works** in accordance with the **Contract**.

**3.2** All work carried out by the **Contractor** shall be carried out with sound workmanship and materials, safely and in accordance with good engineering practice and **Legislation** and shall be to the reasonable satisfaction of the **Project Manager**.

**3.3** The **Contractor** shall set out the **Plant** by reference to points, lines and levels of reference provided to him by the **Project Manager**.

**3.4** The **Plant** as completed by the **Contractor** shall comply with the **Contract** and shall be in every respect fit for the purpose for which it is intended as defined in the **Specification** or in any other provision of the **Contract**.

**3.5** If at any time during the performance of the **Contract** the **Contractor** is of the opinion that a change to the **Works** or the design or method of operation of the **Plant**:

(a) is necessary to eliminate a potential defect in the **Plant** or a specific hazard to any person or party in the performance of the **Works** or in the operation of the **Plant** which has occurred or would otherwise occur; or
(b) would improve operating or life cycle costs of the **Plant**; or
(c) would otherwise be beneficial to the **Purchaser**;

the **Contractor** shall promptly notify the **Project Manager** accordingly.

**3.6** The **Contractor** shall at all times have and maintain adequate resources available for the proper and timely carrying out of the **Works**, including financial resources, and competent, appropriately qualified and experienced personnel.

**3.7** Unless otherwise agreed, the **Contractor** shall at intervals of not more than one calendar month report to the **Project Manager** in accordance with **Schedule 21** (Reports and records) on the progress of the **Works**, supporting his reports with appropriate documents including any proposed revisions to the **Approved Programme**.

The **Contractor** shall provide and maintain in accordance with Sub-clause 40.1 records as required by **Schedule 21**.

**3.8** The **Contractor** shall maintain, and cause **Subcontractors** to maintain, a quality assurance system as described in **Schedule 6** (Quality assurance and validation). The existence of such a quality assurance system shall not relieve the **Contractor** from any of his other duties, obligations or liabilities under the **Contract**. The **Contractor** shall also prepare and implement a validation plan, if such a requirement is included in **Schedule 6**.

**3.9** The **Contractor** shall provide training as described in **Schedule 9** (Training by Contractor).

# 4. Purchaser's responsibilities

**4.1** The **Purchaser**, through the **Project Manager**, shall provide the **Contractor** with the **Documentation** described in **Schedule 3** (Responsibilities of Purchaser) and with all such further information and **Decisions** as are referred to in the **Contract** or are necessary to be provided by the **Purchaser** to permit the **Contractor** to carry out and complete the **Works** at the times specified in the **Contract**. If no such times are specified, the **Purchaser** shall make such provision within a reasonable time with regard to any date stated in **Schedule 11** (Times of completion), the **Approved Programme**, the actual progress of the **Works** and all other relevant circumstances.

**4.2** If the **Contract** provides in **Schedule 3** or any **Special Condition** that the **Purchaser** shall carry out any work or provide any materials, facilities or services which are necessary to permit the carrying out and completion of the **Works** by the **Contractor** then, subject to any express descriptions thereof in the **Contract**, such work, materials, facilities or services shall be carried out and provided:

(a) with sound workmanship and materials, safely and in accordance with good engineering practice and **Legislation**;
(b) in a manner consistent with the proper carrying out and completion of the **Works** in accordance with the **Contract**;
(c) at the times specified in the **Contract**, or if no such times are specified, at reasonable times having regard to any date or period stated in **Schedule 11**, the **Approved Programme**, the actual progress of the **Works** and all other relevant circumstances.

**4.3** If the **Purchaser** will incur costs which are to be reimbursed by the **Contractor** under the **Contract**, the **Purchaser** shall notify the **Contractor** before incurring such costs and shall keep adequate contemporary records of the work, materials and resources involved. Such records shall be open to inspection by the **Contract Manager** at all reasonable times.

**4.4** Subject to any other provision of the **Contract** if the **Contractor** incurs additional **Cost** as a result of any breach of contract by the **Purchaser** the **Contractor** shall be entitled to an addition to the **Contract Price** determined in accordance with and subject to Clause 19 (Claims).

**4.5** The **Purchaser** shall provide personnel with adequate skills and experience:

(a) to carry out their duties under the **Contract**; and
(b) to operate and maintain the **Plant** from the date of the **Take Over Certificate**, and shall make such personnel available for training in accordance with **Schedule 9** (Training by Contractor).

**4.6** The **Purchaser** shall ensure that his and the **Project Manager's** personnel and anybody working under their control shall at all times when at the **Site** be familiar with and comply strictly with the requirements stated in **Schedules 4** (Health & Safety) and **5** (Environmental protection & waste disposal) and any site safety regulations, safe working procedures and health and safety instructions issued to the **Contractor** from time to time by the **Project Manager** or the **Purchaser** and shall further ensure that where necessary such personnel have undergone suitable safety training and certification.

4.7 The **Contractor** shall not be liable for the consequences of any errors or omissions in or arising from work carried out by the **Purchaser** or on the **Purchaser's** behalf by others (other than in respect of any design or information which the **Contractor** is required to verify in accordance with his obligations under the **Contract**).

## 5. Decisions and Contract co-ordination

5.1 Every **Decision**, objection, proposal, claim or report to be issued, given or made under the **Contract** shall be issued, given or made in writing. If any **Decision**, objection, proposal, claim or report is issued, given or made orally, it shall be confirmed in writing by the issuing party within seven days and shall be effective from the time and date of confirmation.

5.2 Any challenge, where permitted by the **Contract**, to a **Decision**, objection, proposal, claim or report shall be supported by a statement and a summary of material facts upon which it relies and shall unless otherwise provided be made within fourteen days of the receipt of the said **Decision**, objection, proposal, claim or report.

5.3 Records of meetings signed in accordance with Sub-clause 29.3 shall constitute a report in the context of this Clause 5 and shall have the same effect as a **Decision**.

5.4 In carrying out the **Contract** both parties shall comply with the requirements set out in **Schedule 20** (Contract co-ordination)

## 6. Sufficiency of Contract Price

6.1 Subject to the provisions of the **Contract**, the cost of carrying out the **Works** shall be at the risk of the **Contractor**, who shall be deemed to have obtained all information and taken account of all circumstances which may affect such cost before agreeing to the **Contract Price**.

6.2 The **Purchaser** shall be responsible for the accuracy of any information provided by him or on his behalf. Except where the **Contract** expressly requires the **Contractor** to verify any such information, if any such information proves to be inaccurate and the **Contractor** consequently incurs any increase in the **Cost** of performing his obligations under the **Contract**, the **Contractor** shall be entitled to an addition to the **Contract Price** determined in accordance with and subject to Clause 19 (Claims).

6.3 If during the carrying out of the **Works** the **Contractor** encounters on the **Site** any physical condition which at the date of the **Contractor's** tender as stated in the **Agreement** would not reasonably have been foreseen by a properly qualified and competent contractor having all the information which the **Contractor** then had or could have obtained by a visual inspection of the **Site** or by reasonable enquiry, and if the **Contractor** considers that he will in consequence of such condition incur an increase in the time or **Cost** of performing his obligations under the **Contract**, he shall notify the **Project Manager** within fourteen days of becoming aware of such unforeseen condition and otherwise shall comply with the requirements of Clause 19. Any such notification shall:

(a) expressly state that it is given under this Sub-clause 6.3;
(b) specify the condition encountered;
(c) specify the steps which the **Contractor** is taking or proposing to take to overcome the condition encountered;
(d) contain estimates of the effect on the **Approved Programme** of the condition and the additional **Cost** which the **Contractor** is likely to incur; and
(e) state how the **Contractor** proposes to minimise such effect and additional **Cost** and/or time.

If the **Contractor** is unable to include in such notification any of the information listed in (b) to (e) above, he shall provide it to the **Project Manager** in accordance with Clause 19.

The **Contractor** shall be entitled to an addition to the **Contract Price** determined in accordance with and subject to Clause 19 in respect of the additional reasonably incurred **Cost** of:

(i) complying with any instruction of the **Project Manager**; or
(ii) in the absence of any instruction from the **Project Manager**, doing whatever is reasonably necessary;

in consequence of the conditions specified in a notification issued under this Sub-clause 6.3.

# 7. Statutory and other obligations

**7.1** The **Purchaser** and the **Contractor** shall in all matters relating to the **Contract** comply with all **Legislation**. The **Purchaser** shall obtain all permits required from government and local authority or any other necessary permissions in connection with the use of the **Site** for the construction, operation and maintenance of the **Plant** with the exception of those permissions, if any, specified in **Schedule 1** (Description of the Works) or in any **Special Condition**, the preparation and submission of the applications for which, together with doing whatever is reasonable to obtain such permissions, shall be the responsibility of the **Contractor**.

**7.2** The **Contractor** shall develop, implement and maintain the Health and Safety and Environmental Plans stated in **Schedules 4** (Health & Safety) and **5** (Environmental protection & waste disposal) to take account of the development of the design, construction and maintenance procedures of the **Plant** having particular regard to the health and safety of all personnel involved with construction work on the **Site** and of all personnel who will be involved with the operation and maintenance of the **Plant**, and the impact of the **Plant** and its operation on the environment.

Such requirements shall not release the **Contractor** from any of his legal responsibilities for safety or environmental protection within the **Site**.

**7.3** If after the date of the **Contractor's** tender as stated in the **Agreement**, there shall be enacted or brought into force any **Legislation** which causes an increase or decrease in the **Cost** to the **Contractor** of carrying out any part of the **Works**, either party or the **Project Manager** may give notification within fourteen days of such circumstances and the **Project Manager** shall determine an appropriate addition to or deduction from the **Contract Price** by means of a **Variation Order**. When reasonably requested to do so by the **Project Manager** the **Contractor** shall assist in such determination by the provision to the **Project Manager** of the make-up of the **Contract Price** or of any relevant part thereof.

**7.4** Notwithstanding anything contained in this Clause 7, all taxes, levies, rates, charges, national insurance contributions and the like assessed on the **Contractor** including any changes therein and all taxes, withholdings and the like on or calculated by reference to the **Contractor's** profits or deemed profits shall be borne by the **Contractor**. For the avoidance of doubt this Sub-clause 7.4 does not apply to Value Added Tax, provision for which is made in Sub-clause 41.13.

# 8. Patent and other protected rights

**8.1** The **Contractor** shall be responsible for all fees, royalties and other charges, payable under the terms of any licence or permission obtained by him in respect of:

(a) any design of the **Plant** provided by the **Contractor**;
(b) the manufacture and supply of the **Plant**;
(c) any work done or method employed in the carrying out of the **Works**; and
(d) **Third Party Software**.

**8.2** The **Purchaser** shall be responsible for the payment of all fees, royalties and other charges, if any, that may be payable under the terms of any licence or permission in respect of the operation or use of the **Plant** or any part thereof to the extent that they are not included in the **Contract Price**.

**8.3** If any design of the **Plant** provided by the **Contractor**, or any item of the **Plant** manufactured or supplied by him, or any work done or method employed by him in carrying out the **Works** should infringe or be claimed to infringe any patent, registered design, design right, trade mark, copyright or other intellectual property right protected by law, the **Contractor** shall indemnify the **Purchaser** against all damages, liabilities, claims, costs and expenses that may result from such infringement or claimed infringement, provided always that this indemnity shall not apply to any such patent, registered design, design right, trade mark, copyright or any other property right protected by law first granted, registered or created after the date of the **Contractor's** tender as stated in the **Agreement** nor to any use of the **Plant** otherwise than for the purpose or in the manner indicated by or reasonably to be inferred from the **Contract**. The **Contractor** shall use reasonable endeavours to keep the **Project Manager** informed of any further patent, registered design, design right, trade mark or copyright which may be published, registered or created and which may affect the obligations of the parties hereunder.

**8.4** In the event of any claim being made or proceedings commenced against the **Purchaser** to which the indemnity set out in Sub-clause 8.3 applies, the **Purchaser** shall promptly notify the **Contractor** who shall:

(a) immediately discuss with the **Purchaser**, the action(s) that the **Contractor** intends to take in dealing with such claim and in conducting such proceedings;
(b) deal with such claim and conduct such proceedings in the **Purchaser's** name; and
(c) at all times keep the **Purchaser** fully informed as to his progress in dealing with such claim or conducting such proceedings.

If within twenty-one days of the **Purchaser's** notification the **Contractor** fails to notify the **Purchaser** that he intends to deal with such claim or conduct such proceedings then the **Purchaser** shall be free to deal with such claim or conduct such proceedings on his own behalf. Unless the **Contractor** has failed to notify the **Purchaser** within the period stated above, or fails at any time to comply with the other requirements of this Sub-clause 8.4, the **Purchaser** shall not make any admission prejudicial to such proceedings.

**8.5** If any **Documentation** or instructions provided to the **Contractor** by the **Purchaser** or the **Project Manager** shall cause or be claimed to cause an infringement of any patent, registered design, design right, trade mark, copyright or other intellectual property right protected by law the **Purchaser** shall indemnify the **Contractor** against all damages, liabilities, claims, costs and expenses that may result from such infringement or claimed infringement.

**8.6** In the event of any claim being made or proceedings commenced against the **Contractor** to which the indemnity set out in Sub-clause 8.5 applies, the **Contractor** shall promptly notify the **Purchaser** who shall:

(a) immediately discuss with the **Contractor** the action(s) that the **Purchaser** intends to take in dealing with such claim and in conducting such proceedings;
(b) deal with such claim and conduct such proceedings in the **Contractor's** name; and
(c) at all times keep the **Contractor** fully informed as to his progress in dealing with such claim or conducting such proceedings.

If within twenty-one days of the **Contractor's** notification the **Purchaser** fails to notify the **Contractor** that he intends to deal with such claim or conduct such proceedings then the **Contractor** shall be free to deal with such claim or conduct such proceedings on his own behalf. Unless the **Purchaser** has failed to notify the **Contractor** within the period stated above, or fails at any time to comply with the other requirements of this Sub-clause 8.6, the **Contractor** shall not make any admission prejudicial to such proceedings.

**8.7** The copyright and all other intellectual rights in all **Documentation** provided by the **Contractor** under the **Contract** shall remain vested in the **Contractor** or his

**Subcontractors** as the case may be, subject to any pre-existing rights of the **Purchaser** or any third party.

8.8 The **Purchaser** shall have the right to use such **Documentation** in connection with the design, construction, operation, maintenance and repair of the **Plant**. The **Purchaser** shall have the further right without additional payment to the **Contractor** to use any such **Documentation** for the purpose of making any improvement to or enlargement of the **Plant**, provided that such improvement or enlargement does not result in the construction of any separate or additional production plant or does not result in an increase in the production capacity of the **Plant** of more than twenty-five per cent over the original design.

8.9 Title in the **Applications Software** shall pass to the **Purchaser** upon its installation within the **Plant**. The **Contractor** shall grant to or obtain for the **Purchaser** an irrevocable royalty-free licence to use for the lifetime of the **Plant** all **Third Party Software** and **Contractor's Software** necessary for the operation or maintenance of the **Plant**.

8.10 Within twenty-eight days of the issue of the relevant **Take-Over Certificate**, the **Contractor** shall supply to the **Purchaser** a copy of the **Applications Software**, together with all **Documentation** necessary to allow the **Purchaser** to maintain, modify and compile the **Applications Software** into an executable form in combination with the **Third Party Software** and the **Contractor's Software**, as detailed in the **Specification**.

# 9. Subcontracting and third party rights

9.1 Neither the **Purchaser** nor the **Contractor** shall without the previous consent of the other transfer any benefit or obligation under the **Contract** to any other person in whole or in part, except that the **Contractor** may without such consent transfer the right to receive any money which is or may become due to him under the **Contract**.

9.2 The **Contractor** may not subcontract the whole of the **Works**.

Subject to Clause 10 (Nominated Subcontractors) and **Schedule 7** (Subcontracting), the **Contractor** may subcontract any part of the **Works** as the **Contractor** considers appropriate.

The subcontracting by the **Contractor** of any part of the **Works** shall not relieve the **Contractor** in any way from his obligations under the **Contract**.

9.3 The **Contractor** shall use reasonable endeavours:

(a) to ensure that each of his **Subcontractors** is required to observe equivalent contractual obligations to those set out in these **General Conditions** and in any **Special Condition** and to observe any provision of the **Contract** which apply to subcontracts: and

(b) to include in any subcontract for any work or **Materials** a provision that the **Subcontractor** shall make good any **Defect** in such work or **Materials** notified at any time before the expiry of the applicable **Defects Liability Period** and that the **Purchaser** shall be entitled to enforce such provision directly against the **Subcontractor**.

9.4 No subcontract shall contain any provision that makes payment to the **Subcontractor** by the **Contractor** dependent upon the receipt by the **Contractor** of any payment due under the **Contract**.

9.5 The **Contractor** shall include in any subcontract under which **Confidential Information** referred to in Sub-clause 20.2 is to be disclosed to the **Subcontractor** a provision that the **Subcontractor** shall observe the same obligations of confidentiality as those of the **Contractor**, and that the **Purchaser** shall be entitled to enforce such provision directly against the **Subcontractor**.

9.6 The **Purchaser** agrees that his obligations under Sub-clauses 20.3, 30.11 and 30.14 shall be capable of being enforced directly by any **Subcontractor**.

**9.7** The **Contractor** agrees that his obligations under Sub-clauses 30.10 and 30.13 shall be capable of being enforced directly by the **Project Manager** or any **Affiliate** of the **Purchaser** to whom the relevant indemnity applies.

**9.8** Save as expressly provided in this Clause 9, no person other than a party to this **Contract** shall have any right to enforce any of its terms under the Contracts (Rights of Third Parties) Act 1999. The **Purchaser** and the **Contractor** may by agreement rescind or vary the **Contract** without the consent of any beneficiary who has rights to enforce a term of the **Contract**, even if as a result that beneficiary's rights to enforce a term of the **Contract** will be varied or extinguished.

# 10. Nominated Subcontractors

**10.1** If the **Contract** provides that any work shall be done or **Materials** or services supplied by a **Subcontractor** named in the **Contract** and stated to be a **Subcontractor** nominated by the **Purchaser**, the **Contractor** shall enter into a subcontract in accordance with such nomination. Sub-clauses 10.3, 10.4 and 10.5 shall not apply in respect of that **Subcontractor**.

**10.2** If the **Contract** provides that any work shall be done or **Materials** or services supplied by a **Subcontractor** to be nominated by the **Project Manager**, the **Project Manager** shall make the necessary nomination at a reasonable time having regard to the progress of the **Works** and, subject to Sub-clauses 10.3 and 10.4, the **Contractor** shall enter into a subcontract in accordance with such nomination.

**10.3** If the **Contractor** considers that:

(a) compliance with the nomination received in accordance with Sub-clause 10.2 would prevent the **Contractor** from or hinder him in fulfilling any of his obligations under the **Contract**; or

(b) the nominated subcontractor is unwilling to enter into a subcontract containing obligations on the subcontractor which correspond to those of the **Contractor** under the **Contract** and provide the **Contractor** with reasonable remedies in the event of a breach thereof by such subcontractor; or

(c) the nominated subcontractor is unlikely to be reliable or competent in his performance of the subcontract;

he shall so notify the **Project Manager** with reasons within twenty-eight days of receipt of the nomination and before entering into a subcontract with the nominated subcontractor.

**10.4** Upon receipt of any notification under Sub-clause 10.3 and unless otherwise agreed:

(a) the **Project Manager** shall withdraw the nomination and nominate another subcontractor in accordance with Sub-clause 10.2; or

(b) if the **Contractor** and the **Project Manager** agree terms the **Project Manager** shall issue a **Variation Order** confirming the nomination, in which case the **Contractor** shall enter into a subcontract with the nominated subcontractor on the terms agreed with the **Project Manager**; or

(c) the **Project Manager** shall withdraw the nomination and issue a **Variation Order** omitting the work, **Materials** or services in question, in which case the **Purchaser** may enter into a direct contract with the previously nominated subcontractor.

**10.5** In the event that the **Contractor** does not notify the **Project Manager** under Sub-clause 10.3 above, he shall be obliged to accept the **Project Manager's** nomination.

**10.6** Subject to the **Contractor** using reasonable endeavours to recover from any nominated **Subcontractor** any loss suffered by the **Contractor** as a result of the failure of that nominated **Subcontractor** to perform his obligations under the relevant subcontract, the **Purchaser** shall indemnify the **Contractor** against all damages, liabilities, claims, costs and expenses that the **Contractor** may suffer as a consequence of that failure by the nominated **Subcontractor** provided that:

(a) the **Contractor** shall notify the **Project Manager** within a reasonable time of becoming aware of such failure;
(b) the failure to perform of such **Subcontractor** did not result from any failure of the **Contractor** himself (and not any **Subcontractor**) to exercise the reasonable skill and care to be expected of a properly qualified and competent contractor; and
(c) the **Contractor** shall not take legal proceedings against that nominated **Subcontractor** without the **Purchaser's** agreement to the action.

**10.7** If any payment to a nominated **Subcontractor** has been improperly withheld by the **Contractor** then the **Purchaser** shall be entitled to notify the **Contractor** that he intends to make such payment direct to the **Subcontractor**. If after fourteen days, the payment is still withheld and the **Contractor** has failed to furnish adequate reason why it should be withheld, the **Purchaser** may, except in the event of the **Contractor's** insolvency, make such payment and shall be entitled to recover from the **Contractor** the amount so paid.

**10.8** Any disagreement arising under this Clause 10 which is not settled in accordance with Clause 46 (Disputes) may be referred to an **Expert** in accordance with Clause 48 (Reference to an Expert).

## 11. The Project Manager

**11.1** The **Project Manager** shall have full authority to act on behalf of the **Purchaser** in connection with the **Contract**, except to the extent otherwise stated in Sub-clause 36.5, and:

(a) the **Purchaser** shall cause the **Project Manager** to perform reasonably and in a timely manner every act required under the **Contract** to be performed by the **Project Manager**;
(b) any obligation stated under the **Contract** to be an obligation of the **Project Manager** shall be deemed to be an obligation of the **Purchaser**;
(c) the **Purchaser** shall be responsible for any act, neglect or omission of the **Project Manager** as if it were an act, neglect or omission of the **Purchaser**; and
(d) in all matters where the **Project Manager** is required or authorised under the **Contract** to exercise his discretion or make a judgment or form an opinion he shall do so impartially and to the best of his skill and judgment as a professional engineer.

**11.2** The **Project Manager** may at any time instruct the **Contractor** to carry out the **Works** or any part thereof as the **Project Manager** may require and the **Contractor** shall comply with such instruction within a reasonable period to be agreed between the **Project Manager** and the **Contractor**.

If the **Contractor** is of the opinion that compliance with an instruction of the **Project Manager** would:

(a) constitute a **Variation**; or
(b) prevent him or hinder him in fulfilling any obligation under the **Contract**; or
(c) be a matter described in paragraph (c), (d), (e), (f) or (g) of Sub-clause 16.6;

he shall so notify the **Project Manager** within fourteen days. The **Project Manager** may request the **Contractor** to provide justification for his opinion. The **Contractor** shall be entitled to an addition to the **Contract Price** for preparing the justification, determined in accordance with and subject to Clause 19 (Claims).

If the **Project Manager** accepts the **Contractor's** opinion, he shall withdraw the instruction and if appropriate issue a **Variation Order** under Clause 16 (Variations).

**11.3** The **Purchaser** may from time to time appoint some other individual as **Project Manager** in place of the individual previously so named or appointed and shall notify the **Contractor** without delay.

**11.4** If the individual initially named in the **Contract** is not an employee of the **Purchaser** or any **Affiliate** of the **Purchaser**, the **Purchaser** shall not thereafter have the right to appoint as **Project Manager** any individual who is an employee of the **Purchaser** or any **Affiliate** of the **Purchaser**, unless the **Contractor** consents.

**11.5** The **Project Manager** may notify the **Contractor** of the appointment of any individual to act as a **Project Manager's Representative** and may from time to time by further notification change such appointment. Any such **Project Manager's Representative** shall have authority to reject any designs, workmanship or materials which are not in accordance with the **Contract**, but the **Contractor** may within seven days of such rejection and before complying with it appeal against it to the **Project Manager**. Any such **Project Manager's Representative** shall also have authority to receive on behalf of the **Project Manager** any notification to be given to the **Project Manager** under the **Contract** by the **Contractor**, but the **Contractor** shall also send a copy of any such notification directly to the **Project Manager**.

**11.6** Subject to Sub-clauses 12.5 and 28.4, the **Project Manager** may notify the **Contractor** that he authorises any **Project Manager's Representative** to exercise any of the powers and functions of the **Project Manager** under the **Contract** and may by further notification cancel or modify any such authority, but not with retrospective effect.

**11.7** If in the absence of the necessary authority from the **Project Manager** a **Project Manager's Representative** has given the **Contractor** any **Decision** which should have been given by the **Project Manager**, the **Contractor** may, within seven days of receipt, advise the **Project Manager** of such **Decision**. Unless the **Project Manager** withdraws that **Decision** within seven days of receipt of such advice, the **Decision** shall have effect as if it had been given by the **Project Manager**. In the event that the **Contractor** does not advise the **Project Manager** of any such **Decision** within seven days of receipt, the **Decision** shall be deemed to have been given with the authority of the **Project Manager**.

**11.8** Except as otherwise provided in or under the **Contract**, all communications between the **Purchaser** and the **Contractor** relating to the **Works** shall be between the **Project Manager** and the **Contract Manager**.

## 12. Contract Manager and Contractor's staff

**12.1** The **Contract Manager** shall have full authority to act on the **Contractor's** behalf in connection with the **Contract**. The **Contractor** shall not replace the **Contract Manager** without the consent of the **Project Manager**. The **Contractor** shall at all times until the issue of the last **Final Certificate** ensure that a suitable individual is appointed to act as **Contract Manager**. The **Contract Manager** shall appoint a deputy to act in his place in his absence. The **Contract Manager** shall notify the **Project Manager** of the appointment of such a deputy who shall be deemed to have full authority to act on the **Contractor's** behalf.

**12.2** From the commencement of work at the **Site** until the whole of the **Plant** has been **Accepted** by the **Purchaser**, the **Contractor** shall ensure that a suitable individual is employed at the **Site** as **Site Manager**. The **Contractor** shall notify the **Project Manager** of the individual so appointed as **Site Manager** and shall not make or change such appointment without the prior consent or requirement of the **Project Manager**.

The **Site Manager** shall supervise all work done at the **Site** by the **Contractor** and shall receive on behalf of the **Contractor** all **Decisions** given to the **Contractor** at the **Site** by the **Project Manager** or any **Project Manager's Representative**. The **Site Manager** shall be present at the **Site** during normal working hours except when on leave, sick or absent for reasons connected with the proper performance of the **Contract** or as agreed with the **Project Manager**. Whenever the **Site Manager** is absent from the **Site** during normal working hours, a suitable individual shall be appointed to act as his deputy. The **Site Manager** shall not be employed by the **Contractor** to do work under other contracts without the agreement of the **Project Manager**.

12.3 The **Contractor** shall provide the further supervisory staff specified in **Schedule 8** (Contractor's named personnel). The **Contractor** shall ensure that at all times there are sufficient suitably qualified and experienced staff to supervise all work being done by the **Contractor** at the **Site** and, where required by the **Contract**, to advise and assist the **Purchaser** in starting up the **Plant**, carrying out the performance tests and operating the **Plant** until its **Acceptance**.

12.4 The **Contractor** shall not replace any key personnel named in **Schedule 8** without the consent of the **Project Manager**.

12.5 If the **Project Manager** is of the opinion that the **Site Manager** or any member of the **Contractor's** supervisory staff at the **Site** is incompetent or has misconducted himself or been in material breach of his duties, he may notify the **Contractor** that he requires such person to leave the **Site** with immediate effect. The **Contractor** shall be under an obligation to act on the **Project Manager's** notification forthwith and to replace such person as soon as practicable. Provided it is given in good faith and does not amount to unlawful discrimination, any **Decision** given by the **Project Manager** under this Sub-clause 12.5 shall be final, conclusive and binding. The **Project Manager** shall not delegate the power to give notification under this Sub-clause 12.5. The **Purchaser** shall have no obligation to pay the **Contractor** for replacing the person.

## 13. Times of completion and Approved Programme

13.1 Subject to Clause 14 (Delays), the **Contractor** shall complete the construction of the **Plant** ready for carrying out the **Take Over** procedures on or before the date, or within the period, specified in **Schedule 11** (Times of completion) and shall also complete the construction of any **Section** of the **Plant** and do any other thing in the performance of the **Contract** on or before the dates, or within the periods, specified in **Schedule 11**.

13.2 The **Contractor** shall within the time stated in **Schedule 11** prepare and submit to the **Project Manager** for his approval a programme of work setting out in any manner that the **Project Manager** may reasonably require the sequence in which and dates by which the **Contractor** proposes to perform his obligations under the **Contract** and the date(s) by which the **Contractor** reasonably requires that the **Purchaser** should provide any further **Documentation** or information or take any other action to permit the **Contractor** to perform his obligations. The programme shall take account of all matters which may affect the completion of the **Works**. If the programme submitted by the **Contractor** is consistent with any dates and periods specified in the **Contract** and is otherwise reasonable, the **Project Manager** shall approve it and it shall be the **Approved Programme**.

The **Contractor** shall use reasonable endeavours to perform his obligations under the **Contract** in accordance with the **Approved Programme**.

13.3 If required by the **Project Manager**, the **Contractor** shall submit details of the qualifications and experience of the personnel whom the **Contractor** proposes to employ on the **Works** at various times, and the other resources which he will have available in order to perform his obligations under the **Contract** and in accordance with the **Approved Programme**. The details shall be in such form as required by the **Project Manager**. The **Contractor** shall employ and make available personnel and resources conforming to such proposals.

13.4 If at any time progress falls behind the **Approved Programme**, or it becomes clear that it will fall behind, the **Project Manager** may require the **Contractor** either to take such steps as may be practicable in order to achieve the **Approved Programme** or to revise the **Approved Programme** in the light of the circumstances and to re-submit it to him for his approval. If the **Project Manager** approves the revised programme it shall thereafter be the **Approved Programme**.

13.5 Without prejudice to Sub-clause 13.4, if the **Project Manager** decides that progress by the **Contractor** in carrying out the **Works** is likely to prejudice the **Contractor's** ability to complete the construction of the **Plant**, or any **Section**, in accordance with Sub-clause 13.1, and that this is due to a cause for which the **Contractor** is responsible, the **Project Manager** may notify the **Contractor** to that effect. Following such notification the **Contractor** shall remedy the potential delay at his own cost.

**13.6** If when the **Project Manager** requires the **Contractor** to revise the **Approved Programme** the **Contractor** fails to submit a revised programme within a reasonable time, or if the **Contractor** submits a revised programme which the **Project Manager** is unable to approve for good reason, the **Project Manager** may instruct the **Contractor** to make reasonable revisions to the programme and the **Contractor** shall as soon as possible make such revisions and the revised programme shall thereafter be the **Approved Programme**.

**13.7** The exercise by the **Project Manager** of his powers under Sub-clauses 13.4, 13.5 and 13.6 above shall not change any of the **Contractor's** obligations to the **Purchaser** under the **Contract**.

## 14. Delays

**14.1** If the **Contractor** is delayed in the performance of any of his obligations under the **Contract** by any of the matters specified in Sub-clause 14.4 or by **Force Majeure** he shall notify the **Project Manager** as soon as possible after becoming aware of the delay.

As soon as reasonably possible after that, the **Contractor** shall notify the **Project Manager** of any extension to any date or period specified in the **Contract** for the completion of such obligations as he considers would be fair and reasonable in the circumstances. The **Contractor** shall keep records made at the time of the circumstances, extent and effect of such delay.

**The Project Manager** shall as soon as possible, and in any event within fourteen days of receipt of the **Contractor's** notification as to the extension to any date or period specified in the **Contract** that the **Contractor** considers would be fair and reasonable, notify to the **Contractor** an initial assessment of any extension to any date or period specified in the **Contract** which the **Project Manager** considers appropriate based on the information available to him at that time.

Thereafter the **Project Manager** may from time to time review and notify to the **Contractor** the need for and amount of any extension to any date or period specified in the **Contract** sought by the **Contractor** having regard to all the information available to the **Project Manager** at the time of any review.

The **Project Manager** shall undertake a review having regard to all the information available to him of the need for and amount of any extension sought by the **Contractor** after the later of either the occurrence of any date or the expiration of any period specified in the **Contract** or as applicable the occurrence of any extension of any date or the expiration of any extension of any period stated in the **Contract**.

The **Project Manager** shall undertake a final review having regard to all the information available to him of the need for and amount of any extension sought by the **Contractor** after the **Contractor** has completed his obligations in respect of any date or period specified in the **Contract**.

In no circumstances shall a later review reduce the amount of an extension previously notified to the **Contractor**.

**14.2** In deciding the appropriate extension to any date or period specified in the **Contract** the **Project Manager** shall have regard to the following:

(a) an extension shall only be granted when the matter causing the delay, being a matter which is specified in Sub-clause 14.4 or is an event of **Force Majeure**, affects achieving the completion of the **Plant** or the **Works** or any part of the **Plant** or the **Works** by any date or by the end of any period stated in the **Contract**;

(b) all the information currently available to the **Project Manager**;

(c) the amount of any extension in respect of a particular cause of delay shall be fair and reasonable; and
(d) in the event of extension for a cause which occurs at a time when the **Contractor** is already in delay for which no extension is allowable, the delay accrued prior to the allowable extension shall not be excused.

**14.3** The notification to the **Contractor** of any extension to any date or period, or any change in any extension, shall be by means of a **Variation Order** stating the appropriate extension to the dates or periods stated in the **Contract** and the **Contractor** shall revise the **Approved Programme** and submit it to the **Project Manager** for approval in accordance with Sub-clause 13.4.

**14.4** The matters entitling the **Contractor** to an extension under Sub-clause 14.1 are delays caused by:

(a) the occurrence of conditions or circumstances to which Sub-clause 6.2, 6.3 or 7.3 applies; or
(b) a **Variation** ordered by the **Project Manager**, other than a **Variation** ordered by reason of the **Contractor's** default; or
(c) the giving in accordance with Sub-clause 42.1 of any instruction by the **Project Manager** to suspend any part of the **Works**, except where due to the **Contractor's** default; or
(d) a breach of the **Contract** by the **Purchaser**; or
(e) the failure of any **Subcontractor** nominated in accordance with Clause 10 (Nominated Subcontractors) to perform the **Subcontractor's** obligations despite all due supervision by the **Contractor**; or
(f) the time taken for the **Purchaser** to confirm to the **Contractor** the availability of funds in accordance with Sub-clause 30.8; or
(g) loss or damage to the **Plant**, **Site Materials** and **Temporary Works** if such loss or damage is due to a **Purchaser's Risk**.

**14.5** If the **Purchaser** is delayed by an event of **Force Majeure** in the performance of any of his obligations under the **Contract**, he shall as soon as possible so notify the **Contractor** and the **Project Manager**.

As soon as possible after that the **Purchaser** shall advise the **Contractor** and the **Project Manager** as to the effect of such event of **Force Majeure**, and shall advise the **Contractor** and the **Project Manager** of any change in the effect of such event of **Force Majeure** including the cessation of the event of **Force Majeure**.

**14.6** **'Force Majeure'** shall mean any circumstance beyond the reasonable control of the parties which prevents or impedes the due performance of the **Contract** by either party including the following:

(a) government action or trade embargo; or
(b) war, hostilities or acts of terrorism; or
(c) riot or civil commotion; or
(d) epidemic; or
(e) earthquake, flood, fire or natural physical disaster; or
(f) exceptionally severe weather conditions or the consequences thereof; or
(g) denial of the use of any railway, port, airport, shipping service or other means of public transport; or
(h) industrial disputes, other than any solely confined to the **Contractor** or his **Subcontractors** or their personnel.

The mere shortage of labour, materials, or utilities shall not constitute **Force Majeure** unless caused by circumstances which are themselves **Force Majeure**.

**14.7** Both parties shall at all times use reasonable endeavours to minimise any delay in the performance of their obligations under the **Contract**, whatever may be the cause of such delay.

**14.8** In the event that either party is delayed by **Force Majeure** each party shall bear his own costs arising from such delay.

**14.9** If performance of the **Works** is substantially prevented by **Force Majeure** for a continuous period of one hundred and twenty days, either party may by **Notice** to the other terminate the employment of the **Contractor**. In the event of such termination the rights and obligations of the parties shall be the same as if the **Purchaser** had issued a **Notice of Termination** under Sub-clause 43.1.

## 15. Damages for delay

15.1 If the **Contractor** fails to complete the construction of the **Plant** or any **Section** or to do any other thing in accordance with **Schedule 11** (Times of completion), the **Contractor** shall pay the **Purchaser** liquidated damages as specified in **Schedule 12** (Liquidated damages for delay), but shall have no liability to pay damages in excess of the maximum (if any) stated in **Schedule 12**.

15.2 After liquidated damages for delay have become payable in respect of any part of the **Plant** or **Works**, if the **Project Manager** issues a **Variation Order** or any event occurs which delays the **Contractor** and in the opinion of the **Project Manager** properly entitles the **Contractor** in accordance with Clause 14 (Delays) to an extension of time in respect of such further delay to that part of the **Plant** or **Works**, the **Project Manager** shall immediately so notify the **Contractor** and the **Purchaser**.

**The Purchaser's** further entitlement to liquidated damages in respect of that part of the **Plant** or **Works** shall thereupon be suspended until the **Project Manager** notifies the **Contractor** and **Purchaser** that such further delay has come to an end.

Such suspension shall not invalidate any entitlement to liquidated damages which became payable before the period of further delay started.

## 16. Variations

16.1 A **Variation** shall mean any change from that stated in the **Contract** to the **Plant**, to the **Works**, to the **Site**, to the method of working, or to the sequence or the timing of work. The **Contractor** shall make no **Variation** except as ordered by the **Project Manager**.

16.2 At any time during the performance of the **Contract** the **Project Manager** may order a **Variation** and, subject to this Clause 16, the **Contractor** shall carry it out as if the **Variation** were part of the original **Contract**. The **Variation** shall be valued in accordance with Clause 18 (Valuation of Variations).

16.3 The **Project Manager** may at any time instruct the **Contractor** to prepare or to assist him in the preparation of a potential **Variation** and the **Contractor** shall comply with such instruction and give to the **Project Manager** his recommendations for the form and scope together with an estimate of the **Cost** and impact on the **Approved Programme** of the **Variation** at such time and in such detail as the **Project Manager** shall require.

16.4 Notwithstanding that consequent additions to or deductions from the **Contract Price** remain to be decided or agreed, the **Project Manager** may issue a **Variation Order** in accordance with Sub-clause 16.2 if he considers that delaying the **Variation Order** pending such decision or agreement would unnecessarily prejudice the satisfactory completion of the **Works** or cause avoidable harm to the interests of the **Purchaser**. In all other cases the **Project Manager** shall give the **Contractor** reasonable opportunity to comment upon any **Variation** he proposes to order and shall request the **Contractor** to state his proposal for consequent adjustments to the **Works**. The **Contractor** shall give his proposal within fourteen days of receipt of such request or within such longer period as the **Project Manager** may agree.

16.5 If the **Contractor** is of the opinion that compliance with any **Variation Order** would prevent him from or hinder him in fulfilling any obligation under the **Contract**, he shall so notify the **Project Manager** as soon as reasonably possible and not later than fourteen days from receipt of the **Variation Order** giving reasons why he considers he would be so prevented or hindered. Any **Variation Order** in respect of which such a notification is given by the **Contractor** shall cease to be binding unless and until it is confirmed by the **Project Manager** with or without modifications. The obligations of the **Contractor** shall be as set out in the **Variation Order** or any modification to it.

**16.6** The **Contractor** may object to any **Variation** ordered or proposed if compliance with it would:

(a) when combined with all **Variations** previously ordered, have a value of more than twenty-five per cent of the **Contract Price**; or

(b) in the case of a **Variation Order** issued after the date of a **Take Over Certificate** for all or part of the **Plant**, have a value of more than five per cent of the **Contract Price**; or

(c) cause the **Contractor** to infringe any **Legislation**, including health, safety or environmental legislation; or

(d) cause the **Contractor** to infringe any patent, registered design, design right, copyright or other protected right of any third party; or

(e) require the **Contractor** to act in breach of any legally binding undertaking or agreement with a third party; or

(f) require the **Contractor** to do work or to exercise skills which are not of the kind he undertakes in the ordinary course of his business, unless the **Variation Order** specifies that the work shall be done or those skills shall be provided by a **Subcontractor** in accordance with Clause 10 (Nominated Subcontractors); or

(g) involve the employment or use of resources beyond the current capacity of the **Contractor** where he can demonstrate that such resources are unavailable to him or cannot reasonably be obtained.

Any such objection by the **Contractor** shall be made to the **Project Manager** as soon as reasonably possible and not later than fourteen days from receipt of the **Variation Order**, giving reasons for his objection. In such case the **Project Manager** shall notify the **Contractor** either that he withdraws the **Variation Order** or proposal, or that he does not accept the validity of the **Contractor's** objection. The decision of the **Contractor** not to object to any **Variation** ordered or proposed under this Clause 16 shall not affect his right to object to any subsequent **Variation Order** or proposal.

**16.7** The **Contractor** shall be entitled to an addition to the **Contract Price** determined in accordance with and subject to Clause 19 (Claims) in respect of the **Cost** of preparing a potential **Variation** in accordance with Sub-clause 16.3, of commenting upon a proposed **Variation** in accordance with Sub-clause 16.4 or of responding to a **Variation Order** in accordance with Sub-clause 16.5 and 16.6, whether or not any such potential or proposed **Variation** is ordered or amended or such **Variation Order** is confirmed or withdrawn.

**16.8** The **Contractor** shall establish and maintain contemporary records of the work, **Materials** and resources required to undertake any **Variation** and their **Cost** together with such additional records as the **Project Manager** may reasonably direct prior to or at the time of issuing the **Variation Order**, and all such records shall be open to inspection by the **Project Manager**.

**16.9** Any disagreement arising under:

(a) Sub-clause 16.5 concerning any modification to the obligations of the **Contractor**; or

(b) Sub-clause 16.6 concerning the validity of the **Contractor's** objection to a **Variation Order**; or

(c) Sub-clause 16.7 concerning any **Cost** incurred by the **Contractor**;

which is not settled in accordance with Clause 46 (Disputes) may be referred to an **Expert** in accordance with Clause 48 (Reference to an Expert).

## 17. Variations proposed by Contractor

**17.1** The **Contractor** may at any time during the performance of the **Contract** submit to the **Project Manager** a proposal for a **Variation**. Initially such proposal should be given as a brief outline as soon as the **Contractor** is first aware of the possibility of the **Variation**.

17.2 If the **Project Manager** wishes to give the proposal further consideration he shall agree with the **Contractor** the extent of the detail that he requires the **Contractor** to prepare in substantiating the proposal or in developing full details of the proposed **Variation**, and thereafter the **Contractor** shall undertake such preparatory work. The **Contractor** shall be entitled to an addition to the **Contract Price** determined in accordance with and subject to Clause 19 (Claims) in respect of the **Cost** of any preparatory work undertaken in accordance with this Sub-clause 17.2.

Any disagreement arising under this Sub-clause 17.2 concerning the **Cost** of any such preparatory work which is not settled in accordance with Clause 46 (Disputes) may be referred to an **Expert** in accordance with Clause 48 (Reference to an Expert).

17.3 If the **Project Manager** decides that the proposed **Variation** should be incorporated into the **Works** or the **Approved Programme**, he shall so order and Clause 16 (Variations) shall apply. The **Project Manager** may instead order or propose an alternative **Variation** under Clause 16.

**The Project Manager's Decision** on any proposal for a **Variation** or on a detailed proposed **Variation** under Sub-clauses 17.1 and 17.2 respectively shall be given within fourteen days of receipt of the proposal for a **Variation** or of the detailed proposed **Variation**, or such longer period as the **Project Manager** notifies the **Contractor** within the said fourteen days is reasonably required for the **Project Manager** and the **Purchaser** to consider the proposal for a **Variation** or detailed proposed **Variation**.

Other than in the special circumstances described in Sub-clause 17.4, a **Decision** of the **Project Manager** not to order a **Variation** or propose an alternative **Variation** shall be final, conclusive and binding.

17.4 The special circumstances referred to in Sub-clause 17.3 are those where the **Contractor** has stated in his proposal that the object of the **Variation** is to eliminate a potential defect in the **Plant** or the **Works** or a hazard to any person or property in the performance of the **Works** or in the operation or maintenance of the **Plant** including any breach of a duty imposed by any health and safety or environmental **Legislation**.

Any disagreement arising under this Sub-clause 17.4 concerning the **Project Manager's Decision** not to order a **Variation** or propose an alternative **Variation** in such special circumstances which is not settled in accordance with Clause 46 may be referred to an **Expert** in accordance with Clause 48.

If such potential defect or hazard derives from a failure of the **Contractor** or if it should have been taken into account by the **Contractor** under Clause 6 (Sufficiency of Contract Price), the **Contractor** shall not be entitled to any addition to the **Contract Price** for the cost of implementing a **Variation** to the extent that it derives from the requirement to eliminate the potential defect or hazard.

17.5 If the **Contractor** becomes aware that any item of the **Plant** or any part of the **Works** to be performed by the **Contractor** has been incorrectly specified in the **Contract**, the **Contractor** shall immediately notify the **Project Manager**. The **Project Manager** and the **Contractor** shall as soon as possible meet to consider what action and what **Variation** may be needed.

# 18. Valuation of Variations

18.1 Subject to **Schedule 18** (Valuation of Variations and claims) the amount, if any, to be added to or deducted from the **Contract Price** in respect of any **Variation** shall be such amount as the **Project Manager** and the **Contractor** may agree. In the absence of such agreement, the **Project Manager** shall as soon as practicable decide such amount which shall in all circumstances be reasonable having regard to **Schedule 18**.

**18.2** Where a fixed price quotation for a **Variation** is provided by the **Contractor** and accepted by the **Project Manager** the price stated in the quotation and accepted shall be the addition to or deduction from the **Contract Price** to be made in respect of such **Variation**.

**18.3** Where so agreed the amount to be added to the **Contract Price** in accordance with this Clause 18 shall be the additional **Cost** the **Contractor** properly incurs plus **Profit**.

# 19. Claims

**19.1** If the **Contractor** intends to claim any additional payment which does not arise out of a **Variation**, he shall notify the **Project Manager** of such intention within fourteen days of becoming aware of the event which gives rise to the claim, and shall establish and maintain records relevant to the claim, together with such additional records as the **Project Manager** may direct. All such records shall be open to inspection by the **Project Manager**.

**19.2** The **Contractor** shall as soon as possible thereafter submit his claim to the **Project Manager** supported by a written statement of grounds and a summary of material facts upon which he relies, together with copies of contemporaneous records and any records made in accordance with Sub-clause 19.1. The **Project Manager** may request any reasonable additional information that he considers necessary to evaluate the claim and the **Contractor** shall provide such information as soon as reasonably practicable.

**19.3** The amount, if any, to be added to the **Contract Price** in respect of any claim by the **Contractor** shall be such amount as the **Project Manager** and the **Contractor** may agree. In the absence of such agreement, the **Project Manager** shall as soon as practicable decide such amount which shall be calculated in accordance with the rates and prices contained in **Schedule 18** (Valuation of Variations and claims) so far as these may be applicable (or otherwise it shall be such amount as shall in all the circumstances be reasonable having regard to **Schedule 18**) and having regard to any prejudice caused to the **Purchaser** by any late notice under Sub-clause 19.1. Any such agreement or decision shall be recorded by means of a **Variation Order**.

**19.4** Where it is stated in the **Contract** that an addition to the **Contract Price** is to be determined in accordance with this Clause 19, the **Contractor** shall be entitled to the additional **Cost** he properly incurs plus **Profit**. The **Contractor** shall not be entitled to any **Cost** incurred as a result of any failure on the part of the **Contractor** to comply with any of the provisions of the **Contract** including those of this Clause 19.

**19.5** Any claim by the **Purchaser** made in accordance with Sub-clause 4.3 shall be submitted to the **Contractor** as soon as possible after the work to which the costs relate has been completed and be supported by a written statement of grounds and a summary of material facts upon which the **Purchaser** relies together with copies of records made in accordance with Sub-clause 4.3. The **Purchaser** and the **Contractor** may agree the costs in respect of any claim by the **Purchaser**. In the absence of such agreement, the **Project Manager** shall decide the amount as soon as practicable.

# 20. Confidentiality

**20.1** **'Confidential Information'** shall mean all **Documentation** and other technical or commercial information in any form obtained directly or indirectly from the **Purchaser** by the **Contractor**, or directly or indirectly from the **Contractor** by the **Purchaser**, or which is generated by the **Contractor** or any **Subcontractor** in connection with the **Contract**, whether before or after the date of the **Contract** other than information:

(a) which is or becomes publicly available other than by any unauthorised action of either of the parties to this **Contract**; or

(b) which is or comes into the possession of one party other than in breach of a duty of confidence to the other party; or

(c) which is expressly approved for disclosure by the party to whom the information relates.

**20.2** The **Contractor** shall not, without the previous consent of the **Purchaser**, use, publish or disclose to any person, nor cause nor permit any of his **Affiliates**, personnel or **Subcontractors** to use, publish or disclose any **Confidential Information** obtained from the **Purchaser** other than for the performance of his duties under the **Contract**.

**20.3** Except as permitted in Clause 8 (Patent and other protected rights) the **Purchaser** shall not, without the previous consent of the **Contractor**, use, publish or disclose to any person, nor cause nor permit any of his **Affiliates**, or personnel to use, publish or disclose to any person any **Confidential Information** obtained from the **Contractor** or any **Subcontractor** otherwise than for the design, construction, operation, repair, maintenance, modification, re-use or removal of the **Plant** and associated facilities.

**20.4** The **Contractor** shall not, other than for the purposes of performing the **Works**, take or permit to be taken any photograph or other image of the whole or any part of the **Plant** or any other property of the **Purchaser** or his **Affiliates**, or any physical or virtual model of it, without the prior consent of the **Purchaser**. Any such photograph or other image shall be regarded as **Confidential Information** within the terms of Sub-clause 20.1. No photograph or other image so taken shall be used for the purposes of publicity without the prior consent of the **Purchaser**.

**20.5** This Clause 20 shall survive and remain in full force for a period of ten years following the issue of the last **Final Certificate** under Clause 38 (Final Certificate) or final **Default Certificate** under Clause 44 (Termination for Contractor's default) unless otherwise stated in any **Special Condition**.

# 21. Documentation

**21.1** The **Contractor** shall appoint competent persons who shall be responsible for the checking and approval of all **Documentation** provided by the **Contractor**. No **Documentation** shall be both checked and approved by the same individual. If such persons are not named in the **Contract**, the **Contractor** shall make such appointments and shall notify the **Project Manager** accordingly.

**21.2** The **Contractor** shall submit to the **Project Manager** for his approval the **Documentation** listed in **Schedule 2** (Documentation) under the heading 'Documentation for approval' at the time or times stated in **Schedule 11** (Times of completion) or in the **Approved Programme**.

Within fourteen days of the receipt by him of any **Documentation** for his approval, the **Project Manager** shall return one copy thereof to the **Contractor** with his approval, conditional approval or rejection endorsed on it. If the **Project Manager** fails to take any such action within the said fourteen days, the **Documentation** shall be deemed to have been approved by the **Project Manager** and the **Contractor** shall notify the **Project Manager** accordingly.

**21.3** The **Project Manager's** approval of any **Documentation** submitted to him shall not relieve the **Contractor** of any of his responsibilities under the **Contract**. The **Contractor** shall not depart from any approved **Documentation** unless he has first submitted amended **Documentation** to the **Project Manager** and obtained his approval thereof.

**21.4** The **Project Manager** shall not reject any **Documentation** except on the grounds that the **Documentation** is not in accordance with some express provision of the **Contract**, or that it is not in accordance with good engineering practice or is likely to have an adverse effect on health or safety or on the environment.

In each case where the **Project Manager** rejects any **Documentation** he shall give his reasons and the **Contractor** shall amend and re-submit such amended **Documentation** to the **Project Manager**.

Any disagreement arising under this Sub-clause 21.4 which is not settled in accordance with Clause 46 (Disputes) may be referred to an **Expert** in accordance with Clause 48 (Reference to an Expert).

**21.5** If the **Project Manager** gives conditional approval to any **Documentation** subject to any comment or query he may have, he shall make such comment or query on or attached to such **Documentation**. The **Contractor** may with the **Project Manager's** consent proceed on the basis of the conditional approval.

The **Contractor** shall in any event re-submit the **Documentation** with his response to the comment or query and the **Project Manager** shall in accordance with Sub-clause 21.2 approve such **Documentation** without conditions when it is re-submitted in accordance with the conditional approval.

**21.6** The **Project Manager** shall have the right at any time on reasonable notice to examine any **Documentation** which has been or is being prepared by the **Contractor** or his **Subcontractors** for the purposes of the **Contract** except any **Documentation** of a class or description which **Schedule 2** or any **Special Condition** states shall not be shown to the **Project Manager**.

**21.7** In addition to the **Documentation** to be submitted under Sub-clause 21.2, **Schedule 2** may include a description of other **Documentation** which is to be provided to the **Project Manager**, under a heading 'Documentation for information only'. The **Contractor** shall submit such **Documentation** to the **Project Manager** as soon as each becomes available in the form specified in **Schedule 2**.

**21.8** **Documentation** relating to any item subject to statutory design requirements and/or insurance approval shall be made available to the **Project Manager** or to the **Purchaser** and/or his insurers or their agents at times appropriate for such purposes, taking account of the **Approved Programme**.

**21.9** At a time or within a period (if any) specified in **Schedule 11** and before the **Plant** is **Taken Over** in accordance with Clause 33 (Taking Over), the **Contractor** shall supply the **Documentation** specified in **Schedule 2** under the heading 'Final Documentation and Manuals' for the **Plant** as actually constructed.

**21.10** The **Project Manager** shall review the **Documentation** supplied by the **Contractor** in accordance with Sub-clauses 21.8 and 21.9 for completeness and accuracy, and may require it to be amended by the **Contractor** as necessary.

**21.11** Whenever the **Contractor** makes any change to the **Plant** as described in the **Documentation** provided in accordance with Sub-clause 21.9, he shall, within a period of sixty days, provide the **Purchaser** with new **Documentation** revised to take account of such change.

**21.12** The **Contractor** shall promptly correct any error, discrepancy or omission in any **Documentation** prepared by him or on his behalf. If the need for correction has arisen by reason of any incomplete or inaccurate data, **Documentation** or information provided by the **Purchaser** or the **Project Manager** on which the **Contractor** reasonably relied, the **Contractor** shall be entitled to an addition to the **Contract Price** determined in accordance with and subject to Clause 19 (Claims). Otherwise the work of the **Contractor** in so doing shall be at his own cost.

**21.13** The **Contractor** shall if required by the **Project Manager** correct any error, discrepancy or omission in any **Documentation** provided to him by the **Purchaser** or the **Project Manager** for the purposes of the **Works** and overcome the consequences of such error, discrepancy or omission and shall be entitled to an addition to the **Contract Price** determined in accordance with and subject to Clause 19. However, if the need for correction has arisen by reason of any inaccurate data, **Documentation** or information provided by the **Contractor** and on which the **Purchaser** or the **Project Manager** reasonably relied, the **Contractor** shall himself bear his additional cost.

**21.14** Provided that the **Purchaser** complies with Sub-clause 4.3 and submits a claim in accordance with Sub-clause 19.5, the **Contractor** shall reimburse to the **Purchaser** any cost of abortive activity which the **Purchaser** has incurred in reliance on any **Documentation** which, in accordance with Sub-clauses 21.12 and 21.13, the **Contractor** is to correct at his own cost. The liability of the **Contractor** to make such reimbursements shall not exceed in aggregate one per cent of the **Contract Price**.

## 22. Inspection and pre-installation tests

**22.1** **The Project Manager** shall be entitled at all times to have access to any place where work under the **Contract** is being carried out or **Materials** are being manufactured or fabricated, for the purpose of checking the progress of manufacture or fabrication, inspecting and observing the work or carrying out tests on the work or **Materials**. The **Project Manager** shall give reasonable notice of the access and facilities he requires and the **Contractor** shall secure such access and facilities both at those premises under his own control and those under the control of his **Subcontractors**.

**22.2** The **Contractor** shall arrange and be responsible for all factory and other off-site tests and those, if any, carried out at the **Site** prior to installation, including those listed in **Schedule 13** (Pre-installation tests and procedures), and shall provide the **Project Manager** with copies of the test results in accordance with the quality assurance system described in **Schedule 6** (Quality assurance and validation).

**22.3** Whenever the **Contractor** is ready to conduct a pre-installation test he shall notify the **Project Manager** of the place and time at which he intends to conduct it, the time being not less than ten days later than the date of such notification. The **Contractor** shall conduct the test at the time and place so notified.

If the **Project Manager** does not attend the test, the **Contractor** shall nevertheless be entitled to carry it out in his absence, and copies of the test results shall be deemed to be a correct record of the test provided that the test is carried out and the results are recorded in accordance with the quality assurance system specified in **Schedule 6**.

**22.4** The **Project Manager** may require any additional pre-installation test not described in **Schedule 13** to be conducted for the purpose of deciding whether any work or **Materials** are in accordance with the **Contract**. The **Project Manager** shall notify the **Contractor** of any such requirement by means of a **Variation Order** in sufficient time to enable the **Contractor** to conduct such test without impeding the performance of his other obligations under the **Contract**. Sub-clauses 22.2 and 22.3 shall apply to any such test. Unless the additional test is one that is normally conducted as part of the practice of the place where work is being done or **Materials** are being manufactured or fabricated, the **Contractor** shall be entitled to an addition to the **Contract Price** determined in accordance with and subject to Clause 19 (Claims).

**22.5** The **Contractor** shall provide or shall arrange for the provision of all labour, materials and equipment necessary for the proper carrying out of all pre-installation tests under this Clause 22.

**22.6** If any item of work or **Materials** fails to pass any pre-installation test, the **Contractor** shall either rectify or replace the relevant item and, unless the **Project Manager** dispenses with a repetition of the test, shall repeat the test following a further notification given under Sub-clause 22.3. The period of such notification may however be less than ten days if a lesser period is reasonable in the circumstances.

**22.7** The **Contractor** shall be responsible for the cost of any repeat inspection or test in the event of an inspection or test failure.

**22.8** The rights of the **Project Manager** under this Clause 22 may be exercised by any authorised nominee of the **Project Manager**.

# 23. The Site

**23.1** The **Purchaser** shall give the **Contractor** access to and possession of the **Site** or parts of the **Site**:

(a) by or before the relevant date or dates specified in the **Contract**; or
(b) if no such date or dates are specified, then in accordance with the **Approved Programme**; or
(c) if there is no **Approved Programme**, or if no such date or dates are specified then in reasonable time to permit the **Contractor** to perform his obligations under the **Contract**.

Such access and possession shall continue until the issue of the **Take Over Certificate**.

Following issue of the **Take Over Certificate** and in accordance with the **Contract** the **Purchaser** shall allow such access as is necessary for the **Contractor** to correct **Defects** and to perform his other rights and obligations under the **Contract**.

**23.2** Unless the **Contract** otherwise provides, the **Purchaser** shall arrange that the **Site** includes a suitable right of access for the **Contractor** from a convenient point on the public highway.

The **Contractor** shall be deemed to have satisfied himself as to the suitability and availability of routes to the **Site** and of the power, water and other utilities which he chooses to use. The **Contractor** shall obtain any permission that may be required from the relevant authorities for the use of such route or utility. Except for information provided in accordance with Sub-clause 4.1, the **Purchaser** does not guarantee and makes no representation as to the suitability or availability of any route or utility but shall provide such assistance as the **Contractor** may reasonably require in obtaining any necessary permission for the use of any route or utility.

**23.3** The **Contractor** shall use every reasonable means to prevent any of the roads (including temporary roads), railways (if any), and all associated bridges and other structures connecting with or on routes to the **Site** from being damaged by any traffic (including in particular, any traffic deemed to be 'extraordinary traffic' within the meaning of **Legislation**) generated by the **Contractor** or any **Subcontractor**. In order to avoid unnecessary damage to any such roads, railways, bridges or other structures, the **Contractor** shall select appropriate routes and vehicles or restrict and distribute loads such that the movement to or from the **Site** of **Contractor's Equipment**, **Materials** or plant is limited as far as reasonably possible.

If the **Works** involve the use by the **Contractor** of waterborne transport, this Sub-clause 23.3 shall be construed as though 'road' or 'route' includes a river, canal, lock, dock, sea wall or other structure related to a waterway, berth or port and 'traffic' includes any kind of waterborne craft, and shall have effect accordingly.

**23.4** The **Contractor** shall ensure that every aspect of and all operations connected with the **Works** are carried out so as not to damage or to interfere unnecessarily with any of the operations being carried out by others on or about the **Site** or with the convenience of the public or the access to or to the occupation and use of any public or private property.

**23.5** The **Contractor** shall permit the **Purchaser** and the **Project Manager** and their personnel to enter the **Site** at all reasonable hours for the purpose of inspecting the **Works**, receiving training or carrying out their work whether or not under or in connection with the **Contract**. The **Contractor** shall also permit other contractors or suppliers engaged by the **Purchaser** to enter the **Site** and carry out work or deliver goods at all reasonable hours, but unless the **Contract** expressly provides otherwise, the **Contractor** shall not be required to allow them to have such access or facilities as would significantly interfere with the performance of his obligations under the **Contract**.

**23.6** The **Contractor** shall not permit to enter the **Site** any person whose presence on the **Site** is not necessary for the carrying out of the **Works** or for the carrying out of other work on behalf of the **Purchaser** without the prior consent of the **Purchaser** or the **Project Manager**.

23.7 The **Contractor** shall take full account of and comply with the **Purchaser's** site and security rules for contractors and ensure that his **Subcontractors** do likewise.

## 24. Delivery to Site

24.1 The **Contractor** shall be responsible for the delivery to the **Site** of all **Materials** and **Contractor's Equipment** in accordance with the requirements of the **Contract**. The delivery arrangements for materials to be provided by the **Purchaser** shall be as stated in **Schedule 3** (Responsibilities of Purchaser).

24.2 Unless otherwise agreed with the **Project Manager**, the **Contractor** shall not deliver or cause to be delivered anything to the **Site** until adequate facilities have been provided at the **Site** for its proper unloading and storage.

24.3 If the **Contractor** wishes any **Materials** or **Contractor's Equipment** to be delivered to the **Site** before the time, if any, specified in the **Approved Programme**, he shall obtain the consent of the **Project Manager** before such delivery is made.

24.4 If any **Materials** or **Contractor's Equipment** are to be delivered to the **Site** seven days or more after the time, if any, specified in the **Approved Programme**, the **Contractor** shall draw this to the attention of the **Project Manager** immediately upon his becoming aware of the delay.

24.5 The **Contractor** shall bear the risk and be liable for any loss of or damage to **Materials** until they are delivered to the **Site** irrespective of any fault or negligence of the **Purchaser** or the **Project Manager**.

24.6 If the **Purchaser** free-issues to the **Contractor** any materials for incorporation into the **Plant** and the **Contractor** first takes possession of them at a location other than the **Site**, then the **Contractor** shall bear the risk and be liable for any loss of or damage to any such materials from the time when the **Contractor** first takes possession of them until delivery to the **Site** irrespective of any fault or negligence of the **Purchaser** or the **Project Manager**.

## 25. Ownership of Materials

25.1 **The ownership of** Materials shall pass to the **Purchaser** at whichever is the earlier of the following:

(a) upon delivery to the **Site**; or
(b) when the **Contractor** becomes entitled to the final payment in respect of such **Materials** and such payment has been made.

The **Contractor** shall insert such provisions in subcontracts as will cause this to happen.

25.2 Whenever the ownership of any **Materials** passes to the **Purchaser** prior to delivery to the **Site**, the **Contractor** shall arrange for the **Materials** to be marked as the **Purchaser's** property and shall ensure that they are stored and handled separately from other materials.

25.3 When so requested, the **Contractor** shall provide to the **Project Manager** proof of ownership and its transfer to the **Purchaser**.

## 26. Health, safety, environment and pollution

26.1 The **Purchaser** and the **Contractor** shall comply with all health, safety and environmental **Legislation**.

Without prejudice to the generality of the foregoing obligation, the **Works** shall be carried out in compliance with the Health and Safety at Work Act 1974 (as amended by Schedule 2 to the Consumer Protection Act 1987) and all applicable Regulations made thereunder, including the Construction (Design and Management) Regulations 2007 or any statutory re-enactment or amendment thereof for the time being in force.

**26.2** The **Contractor** shall be responsible for the safety of his own operations and those of any **Subcontractor** and shall ensure that:

(a) a safe working environment is maintained at all times;
(b) all persons on the **Site**, whether or not employed by the **Contractor** or any **Subcontractor**, are properly protected from risk of injury and danger to health arising out of or in connection with the carrying out of the **Works**;
(c) all property under his control is properly protected from damage or loss;
(d) all construction activities are subject to appropriate risk assessment and that relevant construction method statements are issued;
(e) any hazardous material is identified and suitable safety notices issued; and
(f) any hazardous material for which he is responsible is safely contained or removed from the **Site**.

**26.3** The **Contractor** shall be responsible for the provision of all necessary safety equipment, materials, first aid and emergency arrangements throughout the duration of the construction activities as further described in **Schedule 4** (Health & Safety).

**26.4** The **Contractor** shall ensure that his personnel, **Subcontractors**, and anybody working under their control shall be familiar with, and shall at all times comply strictly with, the requirements stated in **Schedules 4** and **5** (Environmental protection & waste disposal) and any site safety regulations, safe working procedures and health and safety instructions issued to the **Contractor** from time to time by the **Project Manager** or the **Purchaser** and shall further ensure that where necessary his or his **Subcontractor's** personnel have undergone suitable safety training and certification.

**26.5** Without prejudice to the **Contractor's** general responsibility for safety as stated in Sub-clauses 26.1 and 26.2, the **Contractor** shall:

(a) take due notice of all instructions or advice given by the **Purchaser's** designated Health and Safety Supervisor; and
(b) nominate a competent person to be responsible for the proper observance of all safety requirements and procedures during work on the **Site**.

**26.6** The **Contractor** shall ensure that none of his or any **Subcontractor's** personnel performs any activity relating to the performance of the **Works** under the influence of intoxicating liquor, drugs or other similar substances.

**26.7** The **Contractor** shall indemnify the **Purchaser** against any clean-up cost that results from any escape or release of hazardous or polluting material:

(a) from any **Contractor's Equipment** or **Temporary Works**; or
(b) which occurred prior to the issue of the relevant **Take Over Certificate** and was caused by a failure of the **Contractor** to exercise the reasonable skill and care to be expected of a properly qualified and competent contractor; or
(c) which occurred after the issue of the relevant **Take Over Certificate** and before the issue of the relevant **Final Certificate** and was sudden or accidental and was caused by a failure of the **Contractor** to exercise the reasonable skill and care to be expected of a properly qualified and competent contractor.

The **Purchaser** shall indemnify the **Contractor** against any clean-up cost that results from any escape or release of hazardous or polluting material other than as described at (a), (b) or (c) above.

Liability for any loss or damage which may result from the escape or release of hazardous or polluting material other than clean-up costs shall be in accordance with Clause 30 (Care of the Works).

## 27. Site services

**27.1** The **Contractor** shall provide all **Materials**, **Contractor's Equipment**, **Temporary Works**, labour, consumables, services and facilities that may be necessary at the **Site** for the proper carrying out and completion of the **Works** other than those, if any, which the **Contract** requires the **Purchaser** to provide.

**27.2** The facilities and services to be provided by the **Contractor** shall include but not be limited to: temporary roads and parking; temporary offices, stores and warehousing; communications; sanitary and canteen facilities for persons engaged in the **Works**; cleaning and other services; and all necessary or specified safety, security, fire-fighting, first aid and other medical facilities.

**27.3** All temporary buildings, structures, equipment and facilities provided by the **Contractor** at the **Site** shall be properly maintained and cleaned by the **Contractor**, and at all times during the carrying out of the **Works** the **Site** shall be kept by the **Contractor** in a safe, clean and orderly condition and reasonably free from waste materials and rubbish.

**27.4** If the **Contract** states in **Schedule 3** (Responsibilities of Purchaser) or elsewhere that the **Purchaser** shall provide any services or facilities for use by the **Contractor**, then unless otherwise stated, use by the **Contractor** shall be free of charge. Such use by the **Contractor** shall be subject to any condition that the **Project Manager** may reasonably require.

**27.5** Any **Documentation** relating to **Contractor's Equipment**, **Temporary Works** and facilities which the **Contractor** by law or the **Contract** is required to have, make or obtain shall, if required, be produced for inspection by the **Project Manager** or **Purchaser** and/or his insurers or their agents.

**27.6** The **Contractor** shall not remove **Contractor's Equipment** from the **Site** or dismantle or remove **Temporary Works** without the prior consent of the **Project Manager**, but the **Project Manager** shall not refuse consent if the **Contractor's Equipment** or **Temporary Works** are no longer required for the carrying out of the **Works**.

## 28. Site working conditions

**28.1** Subject to Sub-clause 28.3, the **Contractor** shall pay rates of wages and allowances and observe hours and conditions of employment which are not less favourable than those established for the relevant trade or industry by any general agreement or award applying to the place where any work for the **Contract** is being carried out and shall impose the same obligations on his **Subcontractors**.

**28.2** The **Contractor** shall at all times keep the **Project Manager** informed of any matter likely to affect industrial relations at the **Site**.

**28.3** The **Contractor** shall comply with any general agreement made between representatives of workmen and employers who will be engaged in work at the **Site** as to the rates of wages and allowances to be paid to and general working conditions for workmen employed at the **Purchaser's** premises where the **Site** is situated, provided that such agreement was either already in force at the date of the **Contractor's** tender as stated in the **Agreement** or is one to which the **Contractor** is already or becomes a party.

**28.4** The **Contractor** shall ensure the immediate removal from the **Site** of any person, other than his **Site Manager** or a member of his **Site** staff to whom Sub-clause 12.5 applies, whom the **Project Manager** shall by notice require to be removed for reason of misconduct, incompetence or material breach of any of the rules, requirements, regulations, procedures and instructions referred to in Sub-clauses 23.7 and 26.4. Provided it is given in good faith and does not amount to unlawful discrimination, any **Decision** given by the **Project Manager** under this Sub-clause 28.4 shall be final, conclusive and binding. The **Project Manager** shall not delegate the power to give a notice under this Sub-clause 28.4. The **Purchaser** shall have no obligation to pay the **Contractor** for replacing the person.

28.5 Unless otherwise stated in the **Contract**, the **Contractor** shall not, except in an emergency, carry out any work on the **Site** at times outside the normal working hours, if any, specified in **Schedule 1** (Description of the Works) without the prior agreement of the **Project Manager**.

## 29. Meetings

29.1 Meetings to review progress and discuss matters relating to the **Works** shall be held on a regular basis on dates to be fixed by the **Project Manager**. The **Contractor** also may request that a meeting be held at another time if the circumstances require it, and the **Project Manager** shall not unreasonably reject such a request. Meetings shall be attended by the **Project Manager** or the **Project Manager's Representative** and by the **Contract Manager** or **Site Manager** and may be attended by other persons, including representatives of **Subcontractors**, as appropriate.

29.2 The **Project Manager** shall provide a person to record the proceedings of each meeting. Within seven days of the meeting the **Project Manager** shall give two copies of the records signed by him to the **Contract Manager**. If the **Contract Manager** accepts the records as sufficient and accurate, he shall sign one copy and return it to the **Project Manager** within a further seven days. If not, the **Contract Manager** shall agree any modification with the **Project Manager** and the amended copies shall be signed by both of them. If agreement cannot be reached upon any modification to the record of the meeting, the disagreements remaining shall be noted and attached to all copies of the record.

29.3 Only records signed by both the **Contract Manager** and the **Project Manager** shall constitute approved records for the purposes of Sub-clause 5.3.

## 30. Care of the Works

30.1 If the **Contract** provides for the **Plant** to be **Taken Over** by **Sections**, the references to the **Plant** in this Clause 30 shall apply as if a reference to the **Plant** were a reference to a **Section**.

30.2 In this Clause 30, 'Personnel' of a legal entity means any person engaged by and working under the direction and control of that legal entity.

30.3 The **Plant** and **Temporary Works** shall be under the care and control of the **Contractor** until the **Plant** is **Taken Over** by the **Purchaser**. Access to the **Site** and all activities on it under the **Contract** shall, subject to Sub-clause 23.5, be under the control of the **Contractor** until the whole of the **Plant** is **Taken Over** by the **Purchaser**.

30.4 Subject to Sub-clause 30.8, the **Contractor** shall make good any loss or damage to the **Plant**, **Site Materials** and **Temporary Works** that may occur:

(a) before the **Plant** is **Taken Over** by the **Purchaser**; and
(b) after the **Plant** has been **Taken Over** by the **Purchaser** but before the issue of the applicable **Final Certificate**, when the loss or damage results from any cause or operation described in Sub-clause 31.1 (a) or (b).

30.5 The **Contractor** shall have no liability for loss or damage to the **Plant**, **Site Materials** and **Temporary Works** if such loss or damage is due to a **Purchaser's Risk**.

30.6 Subject to Sub-clause 30.5, the **Contractor's** liability for the loss of or damage to the **Plant**, **Site Materials** and **Temporary Works**, including his obligations under Sub-clause 30.4, shall be limited to:

(a) the lesser of:

(i) the **Deductible**; or
(ii) the actual value of each claim to which the **Deductible** shall apply;

plus

(b) any cost that the **Contractor** would have incurred under Clause 37 (Liability for Defects) if replacement or rectification of any **Defect** in the **Plant** and/or **Temporary Works** had been put in hand immediately prior to the physical loss of or damage to the **Plant** and/or **Temporary Works**.

**30.7** The **Purchaser** shall indemnify the **Contractor** in respect of any **Cost** of making good loss of or damage to the **Plant**, **Site Materials** and/or **Temporary Works** in accordance with Sub-clause 30.4 that is in excess of the **Contractor's** liability under Sub-clause 30.6 irrespective of any fault or negligence of the **Contractor** or any **Subcontractor**.

**30.8** To the extent that the **Cost** of making good loss of or damage to the **Plant**, **Site Materials** and/or **Temporary Works** in accordance with Sub-clause 30.4 exceeds the **Contractor's** liability under Sub-clause 30.6 the performance of such obligation shall be conditional upon:

(a) the **Purchaser** confirming that sufficient funds are available to the **Contractor** from the **Purchaser** to implement and complete such work; and
(b) the terms of payment of such funds are agreed between the **Contractor** and the **Purchaser**.

**30.9** The **Contractor** shall make good at the **Contractor's** cost any loss or damage that may occur to the **Documentation** prior to its delivery to the **Purchaser** and/or the **Project Manager**.

**30.10** The **Contractor** shall indemnify the **Purchaser**, any **Affiliate** of the **Purchaser** and the **Project Manager** against all damages, liabilities, claims, costs and expenses in respect of the death, illness or injury to:

(a) Personnel of the **Contractor**; and
(b) Personnel of a **Subcontractor**;

irrespective of any fault or negligence of the **Purchaser**, any **Affiliate** of the **Purchaser**, or the **Project Manager**.

**30.11** The **Purchaser** shall indemnify the **Contractor** and any **Subcontractor** against all damages, liabilities, claims, costs and expenses, in respect of the death, illness or injury to:

(a) Personnel of the **Purchaser**;
(b) Personnel of any **Affiliate** of the **Purchaser**; and
(c) the **Project Manager** and any Personnel of the **Project Manager**;

irrespective of any fault or negligence of the **Contractor** or any **Subcontractor**.

**30.12** Excluding **Temporary Works**, the **Contractor** shall indemnify the **Purchaser** against all losses, damages, liabilities, claims, costs and expenses of the **Contractor** and/or any **Subcontractor** in respect of loss of or damage to:

(a) the **Contractor's Equipment**;
(b) any other property of the **Contractor**; and
(c) any other property of any **Subcontractor**;

irrespective of any fault or negligence of the **Purchaser**.

**30.13** The **Contractor** shall indemnify the **Purchaser** and any **Affiliate** of the **Purchaser** against all losses, damages, liabilities, claims, costs and expenses in respect of loss of or damage to:

(a) the property of the **Purchaser** (other than the **Plant**, **Site Materials** and **Temporary Works**); and
(b) the property of any **Affiliate** of the **Purchaser**;

to the extent caused by the wrongful or negligent act or omission of the **Contractor**, any **Subcontractor**, or the Personnel of the **Contractor** or any **Subcontractor**. The **Contractor's** total aggregate liability to the **Purchaser** and any **Affiliate** of the **Purchaser** shall be limited to the amount as may be stated in the **Agreement** or, if no such sum is stated, to £5,000,000 (five million pounds) in respect of any one incident or series of incidents arising from any one event.

**30.14** Subject to Sub-clause 30.13, the **Purchaser** shall indemnify the **Contractor** and any **Subcontractor** against all damages, liabilities, claims, costs and expenses in respect of loss of or damage to:

(a) the property of the **Purchaser** (other than the **Plant**, **Site Materials** and **Temporary Works**); and
(b) the property of any **Affiliate** of the **Purchaser**;

irrespective of any fault or negligence of the **Contractor** or any **Subcontractor**.

**30.15** Subject to Sub-clause 30.11 and 30.14, the **Contractor** shall indemnify the **Purchaser** against all damages, liabilities, claims, costs and expenses in respect of loss of or damage to:

(a) the property of a third party; and
(b) the death, illness or injury to a third party;

to the extent caused by the wrongful or negligent act or omission of the **Contractor**, any **Subcontractor**, or the Personnel of the **Contractor** or any **Subcontractor**.

# 31. Insurance

**31.1** The **Purchaser** shall provide and maintain a policy or policies of insurance in the joint names of the **Purchaser**, **Project Manager**, **Contractor** and **Subcontractors**, covering all risks (subject to standard exclusions) of loss of or damage to the **Plant**, **Site Materials** and **Temporary Works** to the full cost of their reinstatement (or such other sum as may have been agreed) including cover for loss or damage arising from:

(a) any cause occurring prior to **Take Over**; and
(b) any testing, defect rectification or other activity after **Take Over** carried out or supervised by the **Contractor** or any **Subcontractor** under the **Contract**.

The insurance shall cover physical damage caused by defective design, materials, workmanship, plan or specification, but shall exclude the cost of replacement or rectification which would have been incurred if replacement or rectification of any **Defect** in the **Plant** and/or **Temporary Works** had been put in hand immediately prior to the said physical damage.

The insurance shall contain a waiver of all rights of subrogation against the **Project Manager**, the **Contractor** and **Subcontractors**.

The insurance shall become effective on the commencement of the **Works** and shall remain in place until the issue of the last **Final Certificate**. Unless otherwise agreed the **Purchaser**, as policyholder, shall represent the insured parties in all matters relating to the policy or policies.

**31.2** The **Contractor** shall procure or arrange the provision of an all risk policy or policies of insurance in the joint names of the **Purchaser**, **Project Manager** and the **Contractor** covering the full replacement value of all **Materials** against loss or damage until delivery to the **Site**. If the **Purchaser** free-issues to the **Contractor** any materials for incorporation into the **Plant** and the **Contractor** first takes possession of them at a location other than the **Site**, then the policy shall additionally cover such materials from the time when the **Contractor** first takes possession of them until delivery to the **Site**.

The insurance shall contain a waiver of all rights of subrogation against the **Purchaser** and the **Project Manager**.

**31.3** The **Contractor** shall procure or arrange the provision of a policy or policies of insurance covering loss or damage to the **Contractor's Equipment** to the full replacement value of the **Contractor's Equipment** and becoming effective on the commencement of the **Works** or the date of the **Agreement** (whichever is earlier) and remaining in place until the issue of the last **Final Certificate**.

The insurance shall contain a waiver of all rights of subrogation against the **Purchaser** and the **Project Manager**.

31.4 The **Contractor** shall provide and maintain a policy or policies of third party liability insurance with insurers approved by the **Purchaser** covering liability for death, illness or injury to persons and loss of or damage to property for such sum as the **Contractor** considers appropriate, but in any event not less than £10,000,000 (ten million pounds). The insurance shall contain an indemnity to principals clause. The insurance shall become effective on the commencement of the **Works** or the date of the **Agreement** (whichever is sooner) and shall remain in place until the issue of the last **Final Certificate**.

31.5 The **Contractor** and the **Purchaser** shall each effect and maintain Employer's Liability insurance in accordance with **Legislation**.

31.6 The provisions of this Clause 31 shall not oblige the **Contractor** to take out and maintain a policy of professional indemnity insurance.

31.7 The **Contractor** shall obtain the approval of the **Purchaser** in respect of the insurances to be provided under Sub-clause 31.2 and each party shall provide to the other details of the other insurances he is required to provide by the **Contract** or by **Legislation**. Each party shall also provide details in a timely manner of any additions or restrictions thereto which may be made from time to time. Each party shall provide to the other evidence of the payment of premiums.

If either party fails to produce satisfactory evidence that he has obtained and maintained any insurance required by the **Contract** the other party shall be free, having given seven days notice of his intention, to provide and maintain such insurance and pay the premium as may be necessary for that purpose. The cost of such premium shall be reimbursed by the defaulting party within twenty-eight days of being notified of the amount.

## 32. Completion of construction

32.1 If the **Contract** provides for the completion of construction of the **Plant** to be by **Sections**, Sub-clauses 32.2 to 32.8 shall apply as if a reference to the **Plant** were a reference to a **Section**.

32.2 As soon as the **Plant**, or any part of it, is in the opinion of the **Contractor** substantially complete and ready for inspection he shall so notify the **Project Manager** by means of a draft of a certificate (a **Construction Completion Certificate**), listing the parts of the **Plant** considered to be substantially complete and also the criteria (which may include any test) listed in **Schedule 14** (Criteria for the completion of construction) which apply to such parts of the **Plant**. The **Contractor** shall propose a programme for such inspection and for any tests, commencing not sooner than seven days nor later than fourteen days after the date of the notification unless the **Project Manager** agrees otherwise. If the **Project Manager** requires the **Plant** to comply with any criteria which are not included in **Schedule 14**, he shall order a **Variation**.

32.3 Unless otherwise agreed between the **Project Manager** and the **Contractor**, the **Contractor** shall begin such inspection and tests at the time notified and carry them out in accordance with the quality assurance system specified in **Schedule 6** (Quality assurance and validation). The **Project Manager** shall be entitled to attend and observe them. The **Contractor** shall give the **Project Manager** every reasonable facility to satisfy himself as to the results of the tests.

32.4 Unless the **Contract** otherwise provides, or the **Contractor** and the **Project Manager** otherwise agree, the **Contractor** shall provide all labour, materials and equipment necessary for the carrying out of the inspection and tests.

32.5 If the **Project Manager** does not attend any inspection or test in accordance with Sub-clause 32.3, the **Contractor** shall nevertheless be entitled to carry it out in his absence, and copies of the test results shall be deemed to be a correct record of the test provided that the test is carried out and the results are recorded in accordance with the quality assurance system specified in **Schedule 6**.

32.6 Upon satisfactory completion of any such inspection and tests, the **Project Manager** shall complete and issue to the **Contractor** with a copy to the **Purchaser** the completed **Construction Completion Certificate** confirming that the **Contractor** has demonstrated to the **Project Manager** that the **Plant** or any part of it is substantially complete and in a condition such that any procedures needed to be carried out before the **Plant** is put into operation may be safely carried out. Such **Construction Completion Certificate** may include a note of any minor items requiring completion before the issue of a **Take Over Certificate**.

32.7 If the **Project Manager** is not satisfied that the **Plant** or any part of it listed in a draft **Construction Completion Certificate** in accordance with Sub-clause 32.2 is substantially complete (including complying with the criteria required by **Schedule 14**), he may endorse the draft **Construction Completion Certificate** accordingly, stating in what way the **Plant** or part is incomplete (including not being in accordance with **Schedule 14**). The **Contractor** shall then complete the **Plant** or part as necessary and shall repeat the procedure described in Sub-clause 32.2.

Sub-clause 22.7 shall apply to any repeat inspection or test.

32.8 If in **Schedule 11** (Times of completion), a date is specified for the completion of construction of the **Plant** the **Project Manager** shall, as soon as he has signed all the **Construction Completion Certificates** in accordance with Sub-clause 32.6, issue a **Construction Completion Certificate** for the **Plant**.

32.9 Any disagreement arising under this Clause 32 concerning the issue of a **Construction Completion Certificate** which is not settled in accordance with Clause 46 (Disputes) may be referred to an **Expert** in accordance with Clause 48 (Reference to an Expert).

## 33. Taking Over

33.1 If the **Contract** provides for the **Plant** to be **Taken Over** by **Sections**, Sub-clauses 33.2 to 33.13 shall apply as if a reference to the **Plant** were a reference to a **Section**.

33.2 As soon as the construction of the **Plant** has been demonstrated to be complete in accordance with Clause 32 (Completion of construction) and is, in the opinion of the **Contractor**, ready for carrying out any of the take over procedures (which may include take over tests) specified in **Schedule 15** (Take Over procedures), the **Contractor** shall so notify the **Project Manager** and shall specify a time not sooner than seven days and not later than fourteen days after the date of the notification when the **Contractor** intends to begin to carry out such procedures. If the **Project Manager** requires the **Contractor** to carry out any take over procedures which are not included in **Schedule 15**, he shall order a **Variation**.

33.3 Unless otherwise agreed between the **Project Manager** and the **Contractor**, the **Contractor** shall begin such procedures at the time notified and carry them out in accordance with the quality assurance system specified in **Schedule 6** (Quality assurance and validation). The **Project Manager** shall be entitled to attend and observe them. The **Contractor** shall give the **Project Manager** every reasonable facility to satisfy himself as to the results of any take over tests.

33.4 Unless the **Contract** otherwise provides, or the **Contractor** and the **Project Manager** otherwise agree, the **Contractor** shall provide all labour, materials and equipment necessary for the carrying out of the take over procedures.

33.5 If the **Project Manager** does not attend any take over procedure in accordance with Sub-clause 33.3, the **Contractor** shall nevertheless be entitled to carry it out in his absence, and copies of the test results shall be deemed to be a correct record of the test provided that the test is carried out and the results are recorded in accordance with the quality assurance system specified in **Schedule 6**.

**33.6** If the **Plant** fails to pass a take over test, the **Contractor**, after making such adjustments as he considers necessary, shall repeat such test at a time agreed between the **Contractor** and the **Project Manager**, or notified in accordance with Sub-clause 33.2, and the **Project Manager** shall be entitled to attend and observe the test. If the **Project Manager** is of the opinion that any such adjustment made by the **Contractor** makes it desirable, he may require the **Contractor** to repeat any take over test which has already been successfully carried out, and the **Contractor** shall do so. Subject only to this Sub-clause 33.6 the **Contractor** shall not be bound to repeat for the purposes of this Clause 33 any take over test that has already been successfully carried out.

Sub-clause 22.7 shall apply to any repeat take over test.

**33.7** As soon as any minor items referred to in Sub-clause 32.6 have been completed and, subject to Sub-clause 33.10, all the procedures specified in **Schedule 15** have been successfully carried out, including any which affect the operability or safety of the **Plant**, the **Project Manager** shall issue a certificate (a **Take Over Certificate**) for the **Plant** to the **Contractor** with a copy to the **Purchaser** stating that the **Contractor** has satisfied the requirements of the **Specification** and **Schedule 15**, whereupon the **Plant**, apart from any parts that are excluded from the **Taking Over** by the terms of the Certificate, shall be at the risk of the **Purchaser**. The **Purchaser** shall thereupon be responsible for the care, safety, operation, servicing and maintenance of the **Plant** so certified and shall have the right to operate the **Plant**.

The **Take Over Certificate** may include a list of minor items still to be completed by the **Contractor**, and a list of take over procedures omitted by operation of Sub-clause 33.10.

Any disagreement arising under this Sub-clause 33.7 concerning the issue of a **Take Over Certificate** which is not settled in accordance with the provisions of Clause 46 (Disputes) may be referred to an **Expert** in accordance with Clause 48 (Reference to an Expert).

**33.8** As soon as shall be reasonably possible after the issue of the **Take Over Certificate** the **Contractor** shall complete to the satisfaction of the **Project Manager** all items noted on the **Take Over Certificate** as not being complete at the date of the **Take Over Certificate**. If the **Contractor** fails so to do the **Project Manager** may arrange for such work to be carried out by others in accordance with Sub-clause 37.6.

**33.9** Notwithstanding the failure of the **Plant** to pass any take over test or if certain procedures have not been carried out, the **Project Manager** may with the **Contractor's** consent issue a **Take Over Certificate** in respect of the **Plant**. Such a Certificate may either exclude from **Taking Over** such parts of the **Plant** as are specified, being parts for which the specified procedures have not been carried out, or may specify the take over tests which the **Plant** has failed to pass. In either case the **Contractor** shall carry out any omitted procedure or repeat any test which the **Plant** has not passed as and when required by the **Project Manager** who shall give fourteen days notice of such requirement. The **Contractor** shall become entitled to a separate **Take Over Certificate** as soon as the applicable take over procedures have been completed satisfactorily.

**33.10** If in the opinion of the **Contractor** he has been prevented from carrying out any take over procedure as provided in this Clause 33 by reason of any act or omission of the **Purchaser** or of the **Project Manager**, or of some other contractor employed by the **Purchaser**, then when any other relevant take over procedures have been successfully completed, the **Contractor** may request the **Project Manager** to issue a **Take Over Certificate**. The **Project Manager** shall, if in his opinion such action is reasonable, issue such a Certificate which shall have effect as stated in Sub-clause 33.7. The **Contractor** shall during the **Defects Liability Period** carry out any omitted take over procedure as and when required by the **Project Manager** who shall give fourteen days notice of such requirement.

Any disagreement arising under this Sub-clause 33.10 concerning the **Project Manager's** refusal to issue such a **Take-Over Certificate** which is not settled in accordance with Clause 46 may be referred to an **Expert** in accordance with Clause 48.

**33.11** Any effect on the results of a deferred take over procedure conducted in accordance with Sub-clause 33.9 or 33.10 which can reasonably be shown to be due to the prior use of the **Plant** by the **Purchaser** shall be taken into account in assessing such results.

If the **Contractor** incurs any additional **Cost** as a result of such a deferment, other than any **Cost** resulting from any breach of contract by the **Contractor**, he shall be entitled to an addition to the **Contract Price** determined in accordance with and subject to Clause 19 (Claims).

**33.12** If the **Purchaser** has **Taken Over** the **Plant** as a result of the issue of a **Take Over Certificate** in accordance with Sub-clauses 33.9 or 33.10 and subsequently the **Plant** fails to satisfy the requirements of any relevant take over procedure which had previously failed or been omitted then, subject to the **Contractor** having been given a reasonable opportunity to carry out such take over procedure including making any necessary adjustments to the **Plant**, the **Project Manager** may issue a cancellation certificate to the **Contractor** with a copy to the **Purchaser** stating that any **Acceptance Certificate** which has been deemed to be issued in terms of Sub-clause 36.1 is cancelled. The **Contractor** shall repay all amounts paid to him as a result of such **Acceptance**, and the rights and obligations of the parties shall be as if the **Purchaser** had never **Accepted** the **Plant**.

**33.13** After the issue of a **Take Over Certificate**, and until the **Final Certificate** has been issued, the **Contractor** shall have the right of access, by himself or with any relevant **Subcontractor**, during all reasonable working hours at their own risk and cost to inspect any part of the **Plant** and to take notes of the related working, performance and maintenance records. Access shall only be allowed to the **Contractor's** and any **Subcontractor's** personnel whose names have previously been communicated to the **Project Manager**. Subject to the **Project Manager's** approval, the **Contractor** may at his own risk and cost make any tests which he considers desirable, provided such tests do not interfere unreasonably with the **Purchaser's** use or maintenance of the **Plant**.

## 34. Site clearance

**34.1** Unless otherwise agreed with the **Project Manager**, the **Contractor** shall within fourteen days after the **Purchaser** has **Taken Over** the **Plant** dismantle and remove all **Temporary Works** and remove from the **Site** all waste and debris, **Contractor's Equipment**, including temporary buildings and structures, and other things of his, other than those that are necessary for the proper performance of the **Contractor's** remaining obligations and shall leave the **Site** in an orderly and safe condition. Where the **Plant** is **Taken Over** by **Sections**, the **Contractor** shall remove such things from within the defined boundary of each **Section** as it is **Taken Over**, or if there be no such boundary then from the area about the **Section** as instructed by the **Project Manager**.

**34.2** As soon as the whole of the **Plant** has been **Accepted** the **Contractor** shall remove all his equipment and other possessions from the **Site** other than such things as are then known to be required for making good **Defects** under Clause 37 (Liability for Defects) and any **Materials** which are to be handed over to the **Purchaser** in accordance with Sub-clause 34.5.

**34.3** No **Materials** shall be removed from the **Site** unless:

(a) they are not in accordance with the **Contract** and are being removed for the purpose of:

(i) modification; or
(ii) replacement;

or

(b) the **Project Manager** has previously consented to their removal; or
(c) in the case of bulk materials, they are surplus to the requirements of the **Works**.

34.4 The ownership of all **Materials** removed from the **Site** in accordance with Sub-clause 34.3 (a)(ii), (b) and (c) shall revert to the **Contractor** upon removal. **Materials** falling within the description of Sub-clause 34.3 (a)(i) shall be marked as the **Purchaser's** property and stored and handled separately from other materials.

34.5 The **Contractor** shall remove any surplus **Materials**, except that if the parties shall so agree any surplus **Materials** shall be handed over to the **Purchaser** by the **Contractor** before he leaves the **Site** and an addition to the **Contract Price** shall be determined in accordance with and subject to Clause 19 (Claims) to take account of such **Materials** being additional to the requirements of the **Works**.

If the **Purchaser** shall so require, the **Contractor** shall remove surplus materials which were supplied by the **Purchaser** and dispose of them on behalf of the **Purchaser**. The proceeds of such disposal may be deducted from any sums that would otherwise be payable under the **Contract** by the **Purchaser**, or at the **Purchaser's** option paid by the **Contractor** to the **Purchaser** immediately following such disposal. Otherwise such surplus materials shall be handed over by the **Contractor** before he leaves the **Site**.

## 35. Performance tests

35.1 If under the **Contract** the **Contractor** makes specific guarantees in respect of the performance of the **Plant** as set out in **Schedule 17** (Performance guarantees and damages for failure) this Clause 35 shall apply. Otherwise Sub-clauses 35.2 to 35.18 shall not form part of the **Contract**.

35.2 If the **Contract** provides for the performance of the **Plant** to be tested in **Sections**, Sub-clauses 35.3 to 35.17 shall apply as if a reference to the **Plant** were a reference to a **Section**.

35.3 The performance tests to be carried out on the **Plant** shall be those specified in **Schedule 16** (Performance tests and procedures). If any unspecified test is subsequently proposed, the **Contractor** and the **Project Manager** shall discuss whether such tests should be carried out. If they agree that the test should be carried out, the **Project Manager** shall order the test as a **Variation**.

35.4 The performance tests shall be commenced within the **Performance Test Commencement Period**. If the performance tests have not been commenced within the **Performance Test Commencement Period** the **Project Manager** shall decide, subject to Sub-clause 35.7, an appropriate extension to the **Performance Test Commencement Period**.

If the performance tests have not been commenced within the **Performance Test Commencement Period** because of a delay which is caused by or is the responsibility of the **Purchaser** and if the **Contractor** incurs additional **Cost** as a consequence of such delay the **Contractor** shall be entitled to an addition to the **Contract Price** determined in accordance with and subject to Clause 19 (Claims).

If the performance tests have not been commenced within the **Performance Test Commencement Period** because of a delay which is caused by or is the responsibility of the **Contractor** he shall not be entitled to an addition to the **Contract Price** as a consequence of such delay.

If the performance tests have not been commenced within the **Performance Test Commencement Period** because of a delay which is neither caused by nor is the responsibility of either the **Purchaser** or the **Contractor** the **Contractor** shall not be entitled to an addition to the **Contract Price** as a consequence of such delay.

35.5 The performance tests shall be carried out by the **Purchaser** in the presence of the **Contractor**, unless the **Contractor** fails or declines to attend. The **Purchaser** shall notify the **Contractor** and the **Project Manager** at least fourteen days in advance of his readiness to carry out the performance tests, including a proposal for the time at which the tests should commence. The **Contractor** shall confirm, at least seven days before the time proposed by the **Purchaser**, if he will attend the tests. If the **Contractor** declines or does not attend any performance test the **Purchaser** shall nevertheless be entitled to carry it out in his absence.

The **Performance Test Period** shall commence when the first performance test commences.

**35.6** The performance tests shall be carried out within the **Performance Test Period**.

If the performance tests have not been completed within the **Performance Test Period** because of a delay which is caused by or is the responsibility of the **Purchaser** the **Project Manager** shall, but only if the **Contractor** so agrees and subject to Sub-clause 35.7, decide an appropriate extension to the **Performance Test Period**. If the **Contractor** thereby incurs additional **Cost** the **Contractor** shall be entitled to an addition to the **Contract Price** determined in accordance with and subject to Clause 19.

If the performance tests have not been completed within the **Performance Test Period** because of a delay which is caused by or is the responsibility of the **Contractor** the **Project Manager** shall, but only if the **Purchaser** so agrees and subject to Sub-clause 35.7, decide an appropriate extension to the **Performance Test Period**. The **Contractor** shall not be entitled to an addition to the **Contract Price** under this Sub-clause 35.6 as a consequence of such delay.

If the performance tests have not been completed within the **Performance Test Period** because of a delay which is neither caused by nor is the responsibility of either the **Purchaser** or the **Contractor** the **Project Manager** shall, subject to Sub-clause 35.7, decide an appropriate extension to the **Performance Test Period**. The **Contractor** shall not be entitled to an addition to the **Contract Price** under this Sub-clause 35.6 as a consequence of such delay.

**35.7** Notwithstanding any other provision of the **Contract**, the **Performance Test Commencement Period** and the **Performance Test Period** shall end no later than the end of the **Defects Liability Period**.

**35.8** The performance tests shall be carried out as far as practicable under the conditions detailed in **Schedule 16**. The **Purchaser** shall perform the performance tests using suitably trained and experienced operators and in accordance with the manuals provided by the **Contractor** in accordance with Clause 21 (Documentation) and such other instruction as the **Contractor** may give in the course of carrying out such tests. If any such instruction conflicts in any way with or significantly adds to the manuals, the **Contractor** shall issue it as soon as possible as an amendment to the manuals.

**35.9** Subject to Sub-clause 35.7 every performance test shall be carried out to completion unless an authorised representative of the **Purchaser**, the **Project Manager** or the **Contractor** shall order it to be stopped because its continuance would be unsafe or could damage the **Plant** or other property, or the **Purchaser** stops the test for any other reason.

**35.10** The **Project Manager** shall be responsible for ensuring that data collected during the performance tests are a true and correct record of the tests.

The results of the performance tests shall be compiled and evaluated jointly by the **Purchaser**, the **Project Manager** and the **Contractor**. In the absence of agreement on the results of the performance tests, the **Project Manager** shall determine the results. Any effect on the results of the performance tests caused by prior operation of the **Plant** or arising from a delay, other than a delay which is caused by or is the responsibility of the **Contractor**, shall be taken fully into account in assessing such results.

Any necessary adjustments to the results of the performance tests to take account of actual operating conditions and/or allowances for measurement tolerances shall be made in accordance with **Schedule 16**, or if not stated in **Schedule 16**, in accordance with good engineering practice.

**35.11** If the **Plant** fails to pass any performance test or if any performance test is stopped before its completion the **Purchaser** shall within a reasonable time:

(a) permit access to the **Plant** by the **Contractor** to investigate the causes of any failure to pass a performance test;

(b) permit the **Contractor** to make any adjustment and/or modification to any part of the **Plant**;

(c) shut down any part of the **Plant**, if the **Contractor** reasonably requires it for either of the reasons stated in (a) or (b) above;

(d) restart the **Plant** after completion of any adjustment and/or modification; and/or

(e) repeat any performance test;

provided that the **Performance Test Period** has not expired.

Any investigation into a failure to pass a performance test and any adjustments and modifications shall be made by the **Contractor** as soon as practicable.

The timing of any shutdown referred to in (c) above shall be agreed between the **Contractor** and the **Project Manager**.

The **Contractor** shall, if so required by the **Project Manager**, submit to the **Project Manager** for his approval details of the adjustments and/or modifications which he proposes to make.

The **Contractor** shall make such adjustments and/or modifications at his own cost, provided that if the need for them was caused by the **Purchaser** or by any other contractor employed by the **Purchaser**, the **Contractor** shall be entitled to an addition to the **Contract Price** determined in accordance with and subject to Clause 19.

**35.12** If by the end of the **Performance Test Period** the **Plant** has failed any performance test and provided that the results of the performance tests are within any limit for the application of liquidated damages specified in **Schedule 17**, the **Contractor** shall pay any liquidated damages due to the **Purchaser** in accordance with **Schedule 17**.

**35.13** If by the end of the **Performance Test Period** the results of any performance test remain outside any limit for the application of liquidated damages specified in **Schedule 17**, the **Purchaser** may at his option but subject to Sub-clause 35.7 instruct the **Project Manager** to extend the **Performance Test Period** and instruct the **Contractor** to investigate or to co-operate with the **Project Manager** or others in the investigation of the reasons for the shortfall in performance.

The **Contractor** shall bear his own cost of any such additional work, including any costs associated with any repeat performance test.

**35.14** Subject to Sub-clause 35.13, if by the end of the **Performance Test Period** the results of any performance test are outside any limit for liquidated damages specified in **Schedule 17** the **Project Manager** shall determine the amount of compensation due from the **Contractor** to the **Purchaser** for failure of the **Plant** to comply with a guarantee stated in **Schedule 17**.

**35.15** Subject to Sub-clause 35.16 if by the end of the **Defects Liability Period**:

(a) any performance test has been started but not completed (other than a repeat performance test as described at (b) below); or

(b) any repeat performance test, to be carried out after the **Contractor** has made any adjustment or modification to the **Plant** as a consequence of a failed earlier test, has not been completed (whether or not such repeat test was started);

the **Project Manager** shall determine the amount of compensation (if any) due from the **Contractor** to the **Purchaser** in respect of the performance guarantee associated with any circumstance described at (a) or (b) above having regard to all relevant circumstances and the results (if any) obtained from any completed or partially completed performance test.

**35.16** If for any reason which is not the responsibility of the **Contractor** any performance test has not been started for the first time in sufficient time to allow its completion by the end of the **Defects Liability Period** such performance test shall be deemed to have been passed.

**35.17** Upon the passing of all the performance tests or:

(a) payment or allowance by the **Contractor** of liquidated damages in accordance with Sub-clause 35.12 for failure of the **Plant** to comply with one or more guarantees stated in **Schedule 17**;

(b) payment or allowance by the **Contractor** of compensation in accordance with Sub-clause 35.14 and/or Sub-clause 35.15 for failure of the **Plant** to comply with one or more guarantees stated in **Schedule 17**; and/or

(c) the deemed passing of one or more performance tests in accordance with Sub-clause 35.16 in respect of one or more guarantees stated in **Schedule 17**;

the **Contractor** shall be entitled to the issue of an **Acceptance Certificate** in accordance with Clause 36 (Acceptance).

**35.18** Any disagreement arising under this Clause 35 which is not settled in accordance with Clause 46 (Disputes) may be referred to an **Expert** in accordance with Clause 48 (Reference to an Expert).

# 36. Acceptance

**36.1** If performance tests are to be conducted in accordance with Clause 35 (Performance tests), Sub-clauses 36.2 to 36.5 shall apply. Otherwise any **Take Over Certificate** issued under Clause 33 (Taking Over) shall be deemed to be an **Acceptance Certificate** for the purposes of the **Contract** and Sub-clauses 36.2 to 36.5 shall not apply.

**36.2** If the **Contract** provides for the performance of the **Plant** to be tested in **Sections**, Sub-clauses 36.3 to 36.5 shall apply as if a reference to the **Plant** were a reference to a **Section**.

**36.3** As soon as the **Plant** has passed all the performance tests or otherwise Sub-clause 35.17 has been complied with, the **Project Manager** shall issue to the **Contractor** a certificate (an **'Acceptance Certificate'**) with a copy to the **Purchaser** stating that the **Plant** is **Accepted** by the **Purchaser** as from the date thereof. The **Acceptance Certificate** shall list any known **Defects** which the **Contractor** is bound to make good under Clause 37 (Liability for Defects) and any minor items still remaining to be completed following the issue of a **Take Over Certificate** under Sub-clause 33.7.

**36.4** **Acceptance** of the **Plant** by the **Purchaser** shall not affect the obligations of the **Contractor** under Clause 37.

**36.5** At any time after **Taking Over** the **Plant**, the **Purchaser** may notify the **Contractor** that he elects to **Accept** the **Plant**, expressly stating that his notification is given under this Sub-clause 36.5. The **Project Manager** shall not have authority to give such notification on the **Purchaser's** behalf.

The **Project Manager** shall then issue an **Acceptance Certificate**. The **Acceptance** of the **Plant** under this Sub-clause 36.5 shall constitute a waiver by the **Purchaser** of any remaining obligation of the **Contractor** to ensure that the **Plant** passes the performance tests.

# 37. Liability for Defects

**37.1** If the **Contract** provides for the **Plant** to be **Taken Over** by **Sections**, Sub-clauses 37.2 to 37.12 shall apply as if a reference to the **Plant** were a reference to a **Section**.

**37.2** If at any time before the **Plant** is **Taken Over** in accordance with Clause 33 (Taking Over) or during the **Defects Liability Period**, the **Project Manager**:

(a) decides that any matter is a **Defect**; and
(b) as soon as reasonably practicable notifies the **Contractor** of the particulars of the **Defect**;

the **Contractor** shall as soon as reasonably practicable make good the **Defect** so notified and the **Purchaser** shall so far as may be necessary place the **Plant** at the **Contractor's** disposal for this purpose. The **Contractor** shall, if so required by the **Project Manager**, submit his proposals for making good any **Defect** to the **Project Manager** for his approval.

**37.3** Subject to Sub-clause 37.12, if any **Defect** arises from any breach of the **Contract** by the **Contractor** the **Contractor** shall bear his own cost of making good the **Defect**.

In the case of any other matter made good by the **Contractor**, the work done by the **Contractor** shall be the subject of a **Variation**.

**37.4** If a **Defect** is made good after the issue of a **Take Over Certificate** the **Project Manager** may require the **Contractor** to repeat any appropriate take over test for the purpose of establishing that the **Defect** has been made good. Sub-clause 22.7 shall apply to any such repeat test.

**37.5** If in the course of making good any **Defect** which arises during the **Defects Liability Period** the **Contractor** repairs, replaces or renews any part of the **Plant**, this Clause 37 shall apply to that part of the **Plant** so repaired, replaced or renewed and the **Defects Liability Period** for such part shall commence afresh from the date that the repair, replacement or renewal of such part was completed.

**37.6** If the **Contractor** does not make good within a reasonable time any **Defect** which he is liable to make good under Sub-clause 37.2 or complete any item noted on a **Take Over Certificate** as described in Sub-clause 33.8 then the **Purchaser** may, in addition to any other remedies or relief available to him under the **Contract**, proceed to do the work in such a manner as the **Project Manager** may decide, including the employment of a third party, provided that the **Purchaser** gives at least fourteen days notice of his intention.

**37.7** If the **Purchaser** reasonably requires that any **Defect** notified to the **Contractor** under Sub-clause 37.2 which arises during the **Defects Liability Period** be made good urgently and the **Contractor** is unable or fails to comply within a reasonable time, the **Purchaser** may, in addition to any other remedies or relief available to him under the **Contract** and without further notification, proceed to do the work in such a manner as the **Project Manager** may decide, including the employment of a third party.

**37.8** If the **Purchaser** has made good a **Defect** in accordance with Sub-clause 37.6 or 37.7, the **Contractor** shall reimburse the **Purchaser** his reasonable cost of so doing provided that the **Purchaser** complies with Sub-clause 4.3 and submits a claim in accordance with Sub-clause 19.5. The **Project Manager** and the **Contractor** may agree the amount to be paid by the **Contractor**, or in the absence of agreement the **Project Manager** shall determine such amount as may be reasonable. Such amount shall be:

(a) deducted from any money that would otherwise be payable under the **Contract**; or
(b) paid by the **Contractor** to the **Purchaser**.

Any disagreement arising under this Sub-clause 37.8 which is not settled in accordance with Clause 46 (Disputes) may be referred to an **Expert** in accordance with Clause 48 (Reference to an Expert).

**37.9** If the **Plant** cannot be used because of a **Defect** to which this Clause 37 applies, the **Defects Liability Period** shall be extended by a period equal to the period during which it cannot be used. Similarly the **Defects Liability Period** shall be extended by any period wherein the **Plant** cannot be used by reason of the **Contractor** putting or attempting to put the **Plant** into such condition that it passes any relevant take over procedure under Clause 33 or any relevant performance test under Clause 35 (Performance tests).

**37.10** If for any reason the **Purchaser** does not allow the **Contractor** the necessary access to repair or correct any **Defect**, including permitting the **Contractor** to remove any defective **Materials**, the **Contractor** shall have no obligation to make such repair. However if the **Purchaser** merely defers the timing of such repair or corrective work for his own convenience the **Contractor** shall be entitled to an addition to the **Contract Price** determined in accordance with and subject to Clause 19 (Claims) in respect of any additional **Cost** which the **Contractor** incurs as a consequence of such deferment, irrespective of the underlying liability for the cost of the repair or corrective work. In no circumstance shall the **Purchaser** require the deferment of the commencement of any repair or correction beyond the end of the relevant **Defects Liability Period** or any extension in accordance with this Clause 37.

**37.11** If any part of the **Plant** has a guaranteed working life under normal running conditions stated in **Schedule 10** (Parts with limited working life and spare parts) to be less than the period stated in Sub-clause 1.1 in the definition of the **Defects Liability Period**, then the **Defects Liability Period** for such part shall be the guaranteed working life so stated.

**37.12** If in making good **Defects** which appear in the **Plant** after the date of the relevant **Acceptance Certificate** the **Contractor** incurs any **Cost** in excess of the amount stated in the **Agreement** he shall be entitled to an addition to the **Contract Price** equal to such excess provided that the **Contractor** submits a claim in accordance with and subject to Clause 19.

## 38. Final Certificate

**38.1** Subject to Sub-clause 38.2, as soon as the **Defects Liability Period** for the **Plant** or **Section** has expired and the **Contractor** has made good in accordance with Clause 37 (Liability for Defects) all **Defects** that have within such period appeared in the **Plant** or **Section** and an **Acceptance Certificate** has been issued in respect of the **Plant** or **Section** in accordance with Clause 36 (Acceptance) the **Project Manager** shall issue a certificate (a **'Final Certificate'**) to the **Contractor** with a copy to the **Purchaser** stating that the **Plant** or **Section** and any related work have finally been completed and the date of that completion.

**38.2** If Clause 37 continues to apply to any part of the **Plant**, the **Project Manager** shall as soon as Sub-clause 38.1 is otherwise satisfied, issue a **Final Certificate** for the remainder of the **Plant** or **Section** in which such part is located, clearly identifying such part as being excluded from the **Final Certificate**. Such part shall then be treated as if it were a separate **Section** and shall be the subject of a separate **Final Certificate**.

**38.3** The **Contractor** shall have no right or obligation to do any further work to any part of the **Plant** after a **Final Certificate** has been issued in respect of that part and that **Final Certificate** shall constitute conclusive evidence for all purposes and in any proceedings whatsoever between the **Purchaser** and the **Contractor** that the **Contractor** has completed that part of the **Plant** and made good all **Defects** therein in all respects in accordance with his obligations under the **Contract**.

Where there is more than one **Final Certificate** the last to be issued shall be identified as being the last **Final Certificate**.

No **Final Certificate** shall be conclusive as stated above if it was procured by any fraud on the part of the **Contractor** or any **Subcontractor**.

**38.4** Save where there is an allegation of fraud, any disagreement arising under this Clause 38 concerning whether or not a **Final Certificate** should have been issued which is not settled in accordance with Clause 46 (Disputes) may be referred to an **Expert** in accordance with Clause 48 (Reference to an Expert).

 

# 39. Contract Price

**39.1** Subject to any other term of the **Contract**, the **Contract Price** shall be the sum stated in the **Agreement**.

**39.2** Any money paid in respect of any indemnity under the **Contract** shall not form an addition to the **Contract Price**, except as may be agreed in accordance with Sub-clause 30.8(b) for any payment under the indemnity set out in Sub-clause 30.7.

# 40. Records and audits

**40.1** The **Contractor's** books and records and related documents including those required under Sub-clauses 3.7, 14.1 and 19.1 and **Schedule 21** (Reports and records) or under any other provision of the **Contract** shall be kept and maintained until:

(a) six years after the issue of the last **Final Certificate**; or
(b) any audit undertaken in accordance with Sub-clause 40.2 is complete; or
(c) any dispute resolution procedure is complete;

whichever is the last to occur.

**40.2** The **Purchaser** and the **Project Manager** shall have the right at any time to carry out audits of such books and records and related documents referred to in Sub-clause 40.1 and to have such audits carried out by an auditing firm appointed by the **Purchaser**. Such audit shall commence no later than three hundred and sixty five days after the date of the last **Final Certificate**. It shall not extend to the make-up of any fixed or unit rate or price.

# 41. Payment

**41.1** The **Purchaser** shall pay the **Contractor** the **Contract Price** in instalments as provided in **Schedule 19** (Contract Price and payment).

**41.2** If any instalment is to be paid upon the completion of a specified task or milestone, the **Contractor** shall only be entitled to apply for payment for that instalment when he can provide evidence of completion of the task or milestone as stated in **Schedule 19**.

**41.3** The **Contractor** shall submit a request for payment to the **Project Manager** at intervals of not less than one calendar month showing:

(a) the **Contractor's** assessment of the amount to be paid for **Works** carried out up to the end of the period for which it is submitted, together with any other scheduled payment as may have become payable;

plus

(b) the amounts to which the **Contractor** considers himself entitled in connection with all other matters for which provision is made under the **Contract**.

less

(c) the total of all sums previously certified by the **Project Manager** for payment.

The **Contractor's** requests for payment shall be supported by all relevant documentary evidence appropriately itemised.

The **Contractor's** final request for payment shall state that it is his final request for payment.

**41.4** Within fourteen days of the receipt of an interim request for payment, or in the case of the final request for payment within fifty-six days of its receipt, the **Project Manager** shall issue a certificate to the **Contractor** and the **Purchaser** for the instalment to which the request for payment relates.The certificate shall show the sum which the **Project Manager** considers to be due at the payment due date determined in accordance with Sub-clause 41.5, and the basis on which it has been calculated. The total certified shall comprise all sums listed in the **Contractor's** statement which, in the opinion of the **Project Manager**, are properly payable under the **Contract** and shall show separately any elements within the sums certified in respect of nominated **Subcontractors**. The **Project Manager** may in any certificate delete, correct or modify any sum previously certified by him as he shall consider proper.

**41.5** Any sum certified and notified as due in accordance with Sub-clause 41.4 shall be due for payment fourteen days (or fifty-six days in the case of the final request for payment) after the date of receipt by the **Project Manager** of the **Contractor's** request for payment in accordance with Sub-clause 41.3. The **Purchaser** shall pay the amount so notified by a date (the **Final Date for Payment**) which shall be twenty-eight days (or seventy days in the case of the final request for payment) after the date of receipt by the **Project Manager** of the **Contractor's** request for payment in accordance with Sub-clause 41.3.

**41.6** If for any reason the **Purchaser**, or the **Project Manager** on his behalf, fails to notify the sum due in accordance with Sub-clause 41.4 by the payment due date determined in accordance with Sub-clause 41.5 the sum notified by the **Contractor** in his request for payment in accordance with Sub-clause 41.3 shall be due for payment by the **Final Date for Payment**.

**41.7** If the **Purchaser** intends to pay less than the sum due in accordance with Sub-clause 41.4 or 41.6 for any reason including any sum that may be due from the **Contractor** to the **Purchaser** under the **Contract** or any sum not payable in accordance with Sub-clause 44.5, the **Purchaser** shall notify the **Contractor** not later than one day before the **Final Date for Payment**, specifying the amount he considers to be due on the date the notice is given and the basis on which that sum is calculated.

**41.8** If the **Purchaser** does not make payment in full by the **Final Date for Payment** of the amount of an instalment:

(a) in accordance with any notice issued under Sub-clause 41.7; or
(b) certified and notified in accordance with Sub-clause 41.4; or
(c) where applicable in accordance with Sub-clause 41.6;

or if either party does not make any payment due under any other provision of the **Contract** by the **Final Date for Payment** the amount not paid shall bear interest compounded daily from the **Final Date for Payment** until the amount not paid is received by the other party at an annual rate which is two per cent above the **Agreed Rate** for the first month of delay. The annual rate of interest shall be increased by a further two per cent at the end of each further month of delay, up to the end of the third month. If the delay exceeds three months the annual rate of interest shall thereafter be ten per cent above the **Agreed Rate**.

**41.9** If the **Purchaser** does not make payment in full by the **Final Date for Payment** of the amount of an instalment:

(a) in accordance with any notice issued under Sub-clause 41.7; or
(b) certified and notified in accordance with Sub-clause 41.4; or
(c) where applicable in accordance with Sub-clause 41.6;

the **Contractor** may issue a **Notice** to the **Purchaser** informing him of the **Contractor's** intention to suspend performance of any or all of his obligations under the **Contract**. If such failure shall continue for seven days after the issue of such **Notice**, then at any time thereafter and provided such failure is still continuing, the **Contractor** may suspend further performance of any or all as applicable of his obligations under the **Contract** until payment is made.

The **Project Manager** shall determine by means of a **Variation Order** an extension to the **Approved Programme** and appropriate dates in **Schedule 11** (Times of completion) for the full period of suspension plus any reasonable additional time incurred by the **Contractor** in resuming his obligations and those of his **Subcontractors**. If the **Contractor** incurs additional **Cost** as a result of such suspension and subsequent resumption he shall be entitled to an addition to the **Contract Price** determined in accordance with and subject to Clause 19 (Claims).

**41.10** If the suspension of performance of any, but not all, of the **Contractor's** obligations in accordance with Sub-clause 41.9 continues for a period of sixty days the **Contractor** at any time thereafter may with immediate effect issue a **Notice** to the **Purchaser** and copied to the **Project Manager** that such obligation is excluded from the **Contract**. Thereafter the **Project Manager** shall by means of a **Variation Order** confirm the exclusion together with the applicable deduction from the **Contract Price**.

If the suspension of performance of all of the **Contractor's** obligations in accordance with Sub-clause 41.9 continues for a period of sixty days the **Contractor** at any time thereafter may with immediate effect terminate his employment under the **Contract** by a **Notice** issued to the **Purchaser** and copied to the **Project Manager** and thereupon the rights and obligations of the parties shall be as stated in Clause 43 (Termination by the Purchaser for convenience).

**41.11** If the amount of a payment which is:

(a) notified in accordance with Sub-clause 41.4 (and no notice is given by the **Purchaser** under Sub-clause 41.7 in respect of such payment); or
(b) stated in a notice given by the **Purchaser** under Sub-clause 41.7;

is referred to an adjudicator appointed in accordance with Clause 47 (Adjudication) and if the decision of the adjudicator as to the amount which is to be paid by the **Purchaser** is that more shall be paid than the amount stated as in (a) or (b) as applicable, the additional amount shall be paid not later than:

(i) seven days from the date of the adjudicator's decision; or
(ii) the **Final Date for Payment**;

whichever is the later.

**41.12** If as a result of any audit in accordance with Sub-clause 40.2 or otherwise an error is discovered in the amount paid to the **Contractor** then such error shall be corrected in the next payment due under the **Contract**.

If such error is discovered following the making of the last payment to be certified by the **Project Manager**, the finding of the error shall be notified by one party to the other party as soon as possible. The amount to be paid to or from the **Contractor** shall, if such notification is not challenged within twenty-eight days, become due for payment twenty-eight days after the date of such notification with the **Final Date for Payment** being forty-two days after the date of notification.

If such notification is challenged the parties shall attempt to agree any amount to be paid. If not agreed within fifty-six days of the notification the **Project Manager** shall determine the amount to be paid. The amount to be paid shall be due fourteen days after the date of agreement or determination, with the **Final Date for Payment** twenty-eight days after the date of agreement or determination.

**41.13** The **Contract Price** excludes Value Added Tax and to the extent that the tax is properly chargeable, the **Purchaser** shall pay such tax as an addition to payments otherwise due to the **Contractor**.

## 42. Suspension of the Works

**42.1** The **Project Manager** may instruct the **Contractor** to suspend performance of all or any part of the **Works** for any reason whatsoever and the **Contractor** shall comply with such instruction. The **Project Manager** shall state in any such instruction the reason for the suspension. On receipt of such an instruction, the **Contractor** shall immediately advise the **Project Manager** of any part of the **Works** which needs to be continued to maintain the safety and security of the **Plant** and/or the **Works**.

**42.2** The **Project Manager** may subsequently instruct the **Contractor** to resume some or all of the suspended **Works**. Upon receipt of such an instruction the **Contrac-**

**tor** shall resume such **Works** as are ordered to be resumed as soon as is reasonably practicable.

**42.3** Any extension of the relevant dates or periods stated in **Schedule 11** (Times of completion) notified by the **Project Manager** in accordance with Clause 14 (Delays) shall take account of the additional time required by the **Contractor** in re-establishing his activities and those of his **Subcontractors**.

**42.4** If the **Contractor** incurs any additional **Cost** as a consequence of complying with an instruction to suspend any part of the **Works** given by the **Project Manager**, other than an instruction given by reason of the **Contractor's** default, he shall be entitled to an addition to the **Contract Price** determined in accordance with and subject to Clause 19 (Claims).

**42.5** If the **Contractor's** performance of any of the **Works** is suspended for a continuous period of ninety days, other than by reason of the **Contractor's** default, at any time thereafter and provided that such performance is still suspended, the **Contractor** may issue a **Notice** to the **Purchaser** and to the **Project Manager** requiring that the **Project Manager** shall within fourteen days of such **Notice** either issue an instruction to resume the suspended **Works**, or issue a **Variation Order** excluding that part of the **Works** to which the instruction to suspend applies. If the **Project Manager** shall fail to do so then upon the expiry of such period of notice the employment of the **Contractor** shall unless otherwise agreed be deemed to have been terminated and thereupon the rights and obligations of the parties shall be as stated in Clause 43 (Termination by the Purchaser for convenience).

## 43. Termination by the Purchaser for convenience

**43.1** The **Purchaser** may at any time issue a **Notice** (a **'Notice of Termination'**) to the **Contractor** instructing the **Contractor** to cease the further carrying out of the **Works** for any reason whatsoever other than a reason to which Clause 44 (Termination for Contractor's default) applies.

**43.2** The **Contractor** shall upon receipt of the **Notice of Termination** cease the carrying out of the **Works**, other than such work as the **Project Manager** may instruct. The **Contractor** shall upon completion of such work remove any **Contractor's Equipment** from the **Site** and withdraw from the **Site** leaving it in a safe and tidy condition.

The **Project Manager** shall issue a **Take Over Certificate** in respect of any part of the **Plant** for which a **Take Over Certificate** has not been issued, and shall if the **Contractor** is required to carry out any further work prior to his withdrawal from the **Site** or to provide any **Materials** or **Documentation** applicable to the **Plant** issue any further **Take Over Certificate** as may be appropriate. Thereafter Clauses 37 (Liability for Defects) and 38 (Final Certificate) shall apply.

**43.3** As soon as practicable after receipt of the **Notice of Termination**, the **Contractor** shall if so required by the **Purchaser**:

(a) ensure that his **Subcontractors** cease work in accordance with and subject to Sub-clause 43.2 so far as is possible by the exercise by the **Contractor** of such powers of termination, omission or cancellation as are available to him in the relevant subcontracts; and/or
(b) to the extent permitted by the subcontract, assign any subcontract to the **Purchaser**.

**43.4** Upon receipt of the **Notice of Termination** or on such later date as may be specified in the **Notice of Termination**, the **Contractor** shall transfer to the **Purchaser** to the extent desired by the **Purchaser** all or any rights held by the **Contractor** relating to **Materials**, **Temporary Works** and the **Plant** together with the obligations connected with them. The **Contractor** shall also deliver to the **Purchaser** all erection plans, schedules and **Documentation** and other data prepared by the **Contractor** or **Subcontractors** in connection with the **Works** and all **Confidential Information**

and all other **Documentation** supplied to the **Contractor** by or on behalf of the **Purchaser** in connection with the **Works**.

**43.5** Within ninety days of the **Contractor's** withdrawal from the **Site**, or if the **Contractor** has not entered the **Site** before receipt of the **Notice of Termination** within ninety days of his receipt of the **Notice of Termination**, the **Project Manager** shall, subject to Sub-clause 43.7, issue to the **Purchaser** and to the **Contractor** a certificate (a **'Termination Certificate'**) which shall state:

(a) The total amount (if any) due to the **Contractor** under the **Contract** for the **Works** carried out prior to the receipt by the **Contractor** of the **Notice of Termination**;

(b) any amount due to any third party in respect of which the **Contractor** has (prior to the receipt by him of the **Notice of Termination**) properly and irrevocably entered into a commitment relating directly to the **Contract**; and

(c) the amount of any other **Cost** properly incurred by the **Contractor** in connection with the termination authorised by the **Project Manager**, plus **Profit** thereon;

calculated in accordance with the **Contract**.

The balance, if any, due to the **Contractor** shall be the sum of the amounts referred to in (a), (b) and (c) less that amount already paid to the **Contractor**. If the amount already paid to the **Contractor** exceeds the sum of (a), (b) and (c), the balance shall become due from the **Contractor** to the **Purchaser**.

Any disagreement arising under this Sub-clause 43.5 concerning the amounts payable to the **Contractor** which is not settled in accordance with Clause 46 (Disputes) may be referred to an **Expert** in accordance with Clause 48 (Reference to an Expert).

**43.6** As soon as practicable, the **Contractor** and the **Purchaser** shall provide the **Project Manager** with all such information and documents as he may reasonably require for the purpose of issuing the **Termination Certificate**.

**43.7** If by the expiry of the period specified in Sub-clause 43.5 it is not possible for the **Project Manager** to issue a **Termination Certificate** by reason of any matter which prevents the determination of the amounts referred to in Sub-clause 43.5, the **Project Manager** shall, at the expiration of such period, issue a provisional **Termination Certificate** which shall contain the best estimate that can be made of any amounts referred to in Sub-clause 43.5 and of the resultant balance due. As soon thereafter as the determination of the amounts referred to in Sub-clause 43.5 becomes practicable, the **Project Manager** shall issue a final **Termination Certificate** which shall operate as a correction or adjustment of the provisional **Termination Certificate**.

If in the **Project Manager's** opinion it is appropriate to do so, he may issue more than one provisional **Termination Certificate** prior to the issue of the final **Termination Certificate**.

**43.8** Payment of the balance due under any **Termination Certificate** shall be made between the **Purchaser** and the **Contractor** in accordance with any provisional or final **Termination Certificate** and Clause 41 (Payment) shall apply as appropriate.

**43.9** Nothing in this Clause 43 shall affect the continuing operation of Sub-clauses 9.6, 9.7 and 9.8 and Clauses 8 (Patent and other protected rights), 20 (Confidentiality), 45 (Limitations on liabilities and remedies), 46 (Disputes), 47 (Adjudication), 48 (Reference to an Expert) and 49 (Arbitration) and, to the extent that the **Contractor** has carried out the **Works**, Clauses 30 (Care of the Works), 31 (Insurance) and 37 (Liability for Defects).

## 44. Termination for Contractor's default

**44.1** If the **Contractor** goes into liquidation (other than a voluntary liquidation for the purposes of reconstruction or amalgamation) or has an administration order made against him or carries on his business or any part of it under an administrator or receiver or manager for the benefit of his creditors or any of them, without prejudice to any other rights or remedies, the **Purchaser** may forthwith issue a **Notice** terminating the employment of the **Contractor** under the **Contract**.

**44.2** If the **Contractor** is in default in that he:

(a) without reasonable cause wholly suspends or abandons the carrying out of the **Works** before their completion; or
(b) fails to proceed regularly and diligently with the **Works**; or
(c) commits any other material breach of the **Contract**;

then, without prejudice to any other rights or remedies which the **Purchaser** may possess, the **Purchaser** may issue a **Notice** to the **Contractor** of such default.

If the **Contractor** fails to commence the rectification of the default within a period of fourteen days after receipt of such **Notice**, and/or thereafter fails diligently to pursue the rectification of such default, the **Purchaser** may within fifty-six days of such **Notice** issue a further **Notice** terminating the employment of the **Contractor** under the **Contract**.

The **Purchaser** may at any time give to the **Contractor** a second or subsequent **Notice** of the same default.

**44.3** Upon termination of the **Contractor's** employment under Sub-clause 44.1 or 44.2:

(a) except as the **Project Manager** may direct or permit, the **Contractor** shall forthwith leave the **Site** and shall have no right to re-enter the **Site** or to undertake any work, including the rectification of any **Defect** or to remove any **Contractor's Equipment**, **Temporary Works** or **Materials**;
(b) the **Purchaser** may himself or through others complete the **Works** and he or they may enter upon the **Site** and use the **Contractor's Equipment**, **Temporary Works** and **Materials** and the **Purchaser** shall not be liable to the **Contractor** for any fair wear or tear or accidental damage that may occur to the **Contractor's Equipment**, **Temporary Works** or **Materials**;
(c) the **Contractor** shall deliver promptly to the **Purchaser** all erection plans, schedules and **Documentation** and other data prepared by the **Contractor** or **Subcontractors** in connection with the **Works** and all **Confidential Information** and all other **Documentation** supplied to the **Contractor** by or on behalf of the **Purchaser** in connection with the **Works**; and
(d) the **Contractor** shall, if so required by the **Purchaser** and to the extent permitted by the subcontract, assign any subcontract to the **Purchaser**.

**44.4** If any item of **Contractor's Equipment**, **Temporary Works** or of **Materials** which has not become the property of the **Purchaser** or any other thing is used for the completion of the **Works** in accordance with Sub-clause 44.3(a), when that item or thing is no longer required for the purposes of the **Works**:

(a) the **Purchaser** shall notify the **Contractor** of the availability of such item or thing and shall make such item or thing available for the **Contractor** to collect and remove immediately and the **Contractor** shall so collect and remove such item or thing within twenty-one days of such notification;
(b) if the **Contractor** does not collect and remove the item or thing within a period of twenty-one days of its being made available in accordance with paragraph (a) above, the **Contractor** shall be deemed to have consented irrevocably to the disposal of the item or thing by the **Purchaser** and the **Purchaser** may thereafter dispose of that item or thing as he in his absolute discretion sees fit and any proceeds (less the cost of such disposal) shall be paid to the **Contractor**. The **Contractor** shall be liable to pay to the **Purchaser** those costs of disposal not recovered by the **Purchaser** from any proceeds. The **Project Manager** shall certify the amounts payable to or by the **Contractor** in accordance with this Sub-clause 44.4 and Clause 41 (Payment) shall apply as appropriate save that the **Final Date for Payment** shall be fourteen days after the issue of the **Project Manager's** certificate.

Any money payable under this Sub-clause 44.4 shall not form part of the **Contract Price**.

**44.5** Upon the termination of the **Contractor's** employment under this Clause 44 the **Purchaser** shall have no obligation to make any further payment, whether or not such payment shall have been certified by the **Project Manager** prior to termination, other than as may be certified in accordance with Sub-clause 44.6, 44.7 or 44.9.

The **Purchaser** shall issue any necessary notice under Sub-clause 41.7.

**44.6** Within ninety days of the completion of the **Works** in accordance with Sub-clause 44.3 (b) (including all testing and remedying of **Defects**, such that the total cost to be incurred by the **Purchaser** has been incurred) the **Project Manager** shall, subject to Sub-clause 44.9, issue to the **Purchaser** and the **Contractor** a certificate (a **'Default Certificate'**) which shall give a full statement of account including:

(a) all sums due to the **Purchaser** from the **Contractor** including any cost incurred by the **Purchaser** in completing the **Works** in accordance with Sub-clause 44.3(b) which is in addition to that which the **Purchaser** would have incurred if the **Contractor** had completed the **Works** in accordance with the **Contract**; and

(b) all sums due to the **Contractor** in respect of work completed by the **Contractor** prior to the termination of his employment other than any such work of a temporary nature necessitated by such termination.

Having allowed for all previous payments made to the **Contractor** and any sum due to the **Purchaser** from the **Contractor**, the **Default Certificate** shall state the balance due to or from the **Contractor**.

**44.7** If within sixty days of the date of the termination of the **Contractor's** employment the **Purchaser** has not taken steps to complete the **Works** and the **Plant** in accordance with Sub-clause 44.3, the **Default Certificate** shall be issued by the **Project Manager** to the **Purchaser** and the **Contractor** within ninety days of such date of the termination. The **Default Certificate** shall state the total amount (if any) due to the **Contractor** under the **Contract** for the **Works** carried out prior to the termination of the **Contractor's** employment less the amount of any additional cost properly incurred by the **Purchaser** for any necessary work authorised by the **Project Manager** in connection with the termination.

Having allowed for all previous payments made to the **Contractor** and any sum due to the **Purchaser** from the **Contractor**, the **Default Certificate** shall state the balance due to or from the **Contractor**.

**44.8** As soon as practicable, the **Contractor** and the **Purchaser** shall provide the **Project Manager** with all such information and documents as he may reasonably require for the purpose of issuing the **Default Certificate**.

**44.9** If by the expiry of the period specified in Sub-clause 44.6 or 44.7 it is not possible for the **Project Manager** to issue a **Default Certificate** by reason of any matter which prevents the determination of the amounts referred to in Sub-clause 44.6 or 44.7, the **Project Manager** shall, at the expiration of such period, issue a provisional **Default Certificate** which shall contain the best estimate that can be made of any amounts referred to in Sub-clause 44.6 or 44.7 and of the resultant balance due. As soon thereafter as the determination of the amounts referred to in Sub-clause 44.6 or 44.7 becomes practicable, the **Project Manager** shall issue a final **Default Certificate** which shall operate as a correction or adjustment of the provisional **Default Certificate**.

If in the **Project Manager's** opinion it is appropriate to do so, he may issue more than one provisional **Default Certificate** prior to the issue of the final **Default Certificate**.

**44.10** Any disagreement arising under this Clause 44 concerning the amounts payable by or to the **Contractor** which is not settled in accordance with the provisions of Clause 46 (Disputes) may be referred to an **Expert** in accordance with Clause 48 (Reference to an Expert).

**44.11** Payment of the balance due under any **Default Certificate** shall be made between the **Purchaser** and the **Contractor** in accordance with any provisional or final **Default Certificate** and Clause 41 shall apply as appropriate.

**44.12** If the issue by the **Purchaser** of any **Notice** terminating or purporting to terminate the employment of the **Contractor** under this Clause 44 is subsequently determined to have been invalid, such **Notice** shall not constitute a repudiation of the **Contract** by the **Purchaser** but shall be deemed to have been a **Notice of Termination** issued in accordance with Sub-clause 43.1, and thereupon the rights and obligations of the parties shall be as stated in Clause 43 (Termination by the Purchaser for convenience) and not as stated in this Clause 44.

**44.13** Nothing in this Clause 44 shall affect the continuing operation of Sub-clauses 9.6, 9.7 and 9.8, Clauses 8 (Patent and other protected rights), 20 (Confidentiality), 45 (Limitations on liabilities and remedies), 46 (Disputes), 47 (Adjudication), 48 (Reference to an Expert) and 49 (Arbitration) and, to the extent that the **Contractor** has carried out the **Works**, Clauses 30 (Care of the Works), 31 (Insurance) and 37 (Liability for Defects).

# 45. Limitations on liabilities and remedies

**45.1** Notwithstanding any other provision of the **Contract** neither the **Contractor** nor the **Purchaser** shall be liable to the other for:

(a) wastage, loss or contamination during its use in the **Plant** of any process consumable which shall be deemed to include feedstocks, chemicals, biochemicals, catalysts and utilities; and
(b) loss or deferment of anticipated or actual profit, loss of revenue, loss of use, loss of production, business interruption or any similar damage or for any consequential or indirect losses of any kind resulting from or arising out of or in connection with the **Works** or the performance of them or any act or omission relating to them however caused;

except in respect of:

(i) recoveries obtained as a result of the insurance provided under Clause 31 (Insurance); or
(ii) any sum included within the liquidated damages for delay under Sub-clause 15.1; or
(iii) any sum included within the liquidated damages for failure to pass performance tests under Sub-clause 35.12; or
(iv) any **Profit** that the **Contractor** is entitled to; or
(v) any such losses suffered by a third party (excluding any **Affiliate** of the **Purchaser** or the **Contractor**) to the extent included in any award for damages to or settlement with the third party in connection with a claim covered by the indemnities under Sub-clauses 8.3, 8.5, 26.7 and 30.15.

**45.2** Except in the case of termination of the **Contractor's** employment under Clause 44 (Termination for Contractor's default) or a repudiation of the **Contract** by either party, the liability of either party to the other arising out of or in connection with the **Contract** or the **Works**, whether by reason of any breach of contract or of statutory duty or tortious or negligent act or omission shall be limited to the damages, remedies and reimbursements expressly provided in the **Contract**. Nothing in the **Contract** shall in any way be interpreted as affecting or limiting any liability which the **Contractor** may have under the Consumer Protection Act 1987 or in respect of personal injury or death caused by the negligence of the **Contractor** (as defined in Section 1 of the Unfair Contract Terms Act 1977).

**45.3** The total aggregate liability of the **Contractor** to the **Purchaser** arising out of or in connection with the **Contract** and the **Works**, other than liability arising under Sub-clauses 8.3 and 26.7 and Clauses 30 (Care of the Works) and 31 (Insurance), shall not exceed the amount stated in the **Agreement**.

For the purposes of this Sub-clause 45.3 any deduction from or abatement of the **Contract Price** to which the **Purchaser** may become entitled as a result of:

(a) a **Defect**; or
(b) any other failure of the **Contractor** to comply with the terms of the **Contract**;

shall be deemed to be a liability of the **Contractor** to the **Purchaser**.

**45.4** Sub-clauses 45.1 and 45.3 shall apply whether or not either party shall be held to have repudiated the **Contract**.

**45.5** Any exclusion or limitation of liability under the **Contract** shall as between the parties exclude or limit such liability in contract, tort/delict including negligence or otherwise.

## 46. Disputes

**46.1** The **Purchaser** and the **Contractor** shall endeavour to avoid the escalation of problems into disputes as defined in Sub-clause 46.4 and to avoid disputes both between themselves and with third parties including **Subcontractors**. The parties acknowledge and agree that this shall not prejudice any right either party may have under Clause 47 to refer any dispute or difference to adjudication at any time, if that party so wishes, notwithstanding the definition of 'dispute' in Sub-clause 46.4.

**46.2** In order to avoid the development of disputes and to facilitate their clear definition and early resolution, the procedures set out in Clauses 46 (Disputes), 47 (Adjudication), 48 (Reference to an Expert) and 49 (Arbitration) shall be applied as appropriate.

**46.3** If the **Contractor** is dissatisfied with any **Decision** or valuation of the **Project Manager**, or of any person to whom the **Project Manager** may have delegated any of his authority or responsibility, or if the **Purchaser** or the **Contractor** is dissatisfied with any other matter arising under or in connection with the **Contract**, either party may at any time refer such dissatisfaction to the **Project Manager** giving full details of the nature of the matter. The **Project Manager** shall give a **Decision** on the matter (giving the reasons for such **Decision**) to the **Purchaser** and the **Contractor** within twenty-eight days of such reference to him.

**46.4** The **Purchaser** and the **Contractor** agree that no matter shall constitute or give rise to a dispute unless it has been referred to the **Project Manager** under Sub-clause 46.3 and:

(a) the **Project Manager** has failed to give his **Decision** on the matter within the prescribed time; or

(b) a **Decision** given within the prescribed time is either unacceptable to the **Purchaser** and/or the **Contractor** or has not been implemented within twenty-one days of the **Decision**;

and, as a consequence, either the **Purchaser** or the **Contractor** has issued to the other a **Notice** (a '**Notice of Dispute**') setting out the nature of the dispute with a copy to the **Project Manager**. For the purposes of the performance of the **Works** and all matters arising out of or in connection with the **Contract**, the word 'dispute' shall be construed in accordance with this Sub-clause 46.4.

**46.5** Notwithstanding the existence of any dispute or any reference to the **Project Manager** under Sub-clause 46.3, the **Purchaser** and the **Contractor** shall continue to perform their obligations under the **Contract**.

**46.6** The parties shall attempt in good faith to negotiate a settlement of any dispute or difference.

**46.7** If a dispute cannot be resolved by negotiation the parties may by agreement refer it to mediation in accordance with the procedures of the Centre for Effective Dispute Resolution (CEDR), the Chartered Institute of Arbitrators (CIArb) or some other appropriate body.

**46.8** No **Decision**, opinion, direction, or valuation given by the **Project Manager** shall disqualify him from being called as a witness and giving evidence before any third party, an **Expert**, adjudicator or arbitrator on any matter whatsoever relating to a dispute.

## 47. Adjudication

47.1 This Clause 47 shall only apply to disputes arising under a construction contract as defined in the Housing Grants, Construction and Regeneration Act 1996, or any amendment or re-enactment thereof.

47.2 Notwithstanding any provision in the **Contract** for a dispute to be referred to an **Expert** in accordance with Clause 48 (Reference to an Expert) or to arbitration in accordance with Clause 49 (Arbitration), either party shall have the right to refer any dispute or difference (including any matter not referred to the **Project Manager** in accordance with Sub-clause 46.3) as to a matter under or in connection with the **Contract** to adjudication and either party may, at any time, issue a **Notice** (a **'Notice of Adjudication'**) to the other stating his intention to do so . The ensuing adjudication shall be conducted in accordance with the edition of the 'Adjudication Rules' published by **IChemE** current at the time of service of the **Notice of Adjudication**.

47.3 Unless the adjudicator has already been appointed, he is to be appointed to a timetable with the object of securing his appointment and referral of the dispute to him within seven days of the service of the **Notice of Adjudication**. The adjudicator shall be appointed in accordance with the Adjudication Rules.

47.4 The adjudicator shall reach his decision within twenty-eight days of referral or such other longer period as may be agreed between the parties after the dispute has been referred.

47.5 The adjudicator may extend the period of twenty-eight days by up to fourteen days with the consent of the party by whom the dispute was referred.

47.6 The adjudicator shall act impartially.

47.7 The adjudicator may take the initiative in ascertaining the facts and the law.

47.8 The decision of the adjudicator shall be binding until the dispute is finally determined by legal proceedings, by arbitration or by agreement.

47.9 The adjudicator may correct his decision so as to remove a clerical or typographical error arising by accident or omission.

47.10 The adjudicator may include in his decision a direction as to the allocation of his fees and expenses as between the parties.

47.11 The adjudicator shall not be liable for anything done or omitted in the discharge or purported discharge of his functions as adjudicator unless the act or omission is in bad faith. Furthermore, any personnel of the adjudicator acting in connection with the carrying out of the adjudication shall be similarly protected from liability.

## 48. Reference to an Expert

48.1 Subject to the provisions of Clause 46 (Disputes), any dispute which in accordance with the **Contract** is a matter which may be referred to an **Expert** in accordance with this Clause 48 shall be so referred upon one party issuing a **Notice** thereof to the other. Furthermore, if the parties so agree any other dispute arising out of or in connection with the **Contract** may be referred to an **Expert**.

Any such reference shall be conducted in accordance with the edition of the 'Rules for Expert Determination' published by **IChemE** current at the time of the **Expert's** appointment unless the parties otherwise agree prior to the appointment of the **Expert**. The **Expert** shall be appointed in accordance with the Rules for Expert Determination.

48.2 The **Expert** shall decide all disputes referred to him as an expert and not as an arbitrator. Any decision of the **Expert** may revise or overrule any **Decision** of the **Project**

**Manager** other than any decision expressly stated in the **Contract** to be final, conclusive and binding.

48.3 The powers of the **Expert** to decide disputes shall include decisions as to factual issues and the interpretation of the **Contract**.

48.4 The **Purchaser** and the **Contractor** shall give the **Expert** every assistance in deciding any dispute referred to him and shall give him access to the **Site** and to their premises and shall provide him with any information and **Documentation** that he may reasonably require.

48.5 The **Expert** shall have power by his decision to fix the reasonable amount of his fees in connection therewith and they shall be borne in equal shares between the **Purchaser** and the **Contractor**.

48.6 If the **Contractor** has complied with any **Decision** of the **Project Manager** which is overruled by the **Expert**, and has thereby suffered or incurred any loss or additional expense in the performance of the **Contract**, then provided that the **Contractor** has notified the **Project Manager** that he disputes such **Decision** before complying with it, the **Expert** may in his discretion award the **Contractor** a reasonable sum in respect of such loss or additional expense, notwithstanding that the overruling of the **Decision** would not otherwise entitle the **Contractor** to any additional payment under the **Contract**.

48.7 The **Purchaser** and the **Contractor** agree to be bound by any decision of the **Expert** under this Clause 48, which decision shall be final, conclusive and binding, and shall comply with any direction given therein and shall not question the correctness of any such decision or direction in any proceedings.

48.8 Neither the **Purchaser** nor the **Contractor** shall be entitled to suspend performance of the **Contract** by reason of the reference of a dispute to an **Expert**.

48.9 Any dispute which is referred to an **Expert** ceases to be referable to arbitration under Clause 49 (Arbitration).

48.10 A limitation period on actions in litigation or arbitration which applies to the **Contract** under the applicable law shall apply also to the issue of a **Notice** to refer a dispute to an **Expert** in accordance with Sub-clause 48.1.

## 49. Arbitration

49.1 Any dispute, other than failure to give effect to a decision of an adjudicator given under Clause 47 (Adjudication) or of an **Expert** given under Clause 48 (Reference to an Expert), which has not been referred to an **Expert** in accordance with Clause 48 shall be referred to a single arbitrator by one party serving a **Notice** (a '**Notice of Arbitration**') on the other.

49.2 The arbitration shall be conducted in accordance with the procedure set out in the edition of the 'Arbitration Rules' published by **IChemE** current at the time of the appointment of the arbitrator. The arbitrator shall be appointed in accordance with the Arbitration Rules.

49.3 The arbitrator shall have full power to open up, review, revise or overrule any **Decision** of the **Project Manager**, other than any **Decision** expressly stated in the **Contract** to be final, conclusive and binding.

Neither party shall be limited in arbitration to the evidence or arguments put to the **Project Manager** or to an adjudicator in accordance with Sub-clause 46.3 and Clause 47 respectively.

49.4 The **Purchaser** and the **Contractor** agree to be bound by any decision, award or ruling of an arbitrator under this Clause 49 and shall comply therewith. The **Purcha-**

 

**ser** and the **Contractor** hereby agree to exclude any right to appeal to the Court under the Arbitration Act 1996.

**49.5** If the **Contractor** has complied with any **Decision** of the **Project Manager** which is overruled by the arbitrator, and has thereby suffered or incurred any loss or additional expense in the performance of the **Contract**, then provided that the **Contractor** has notified the **Project Manager** that he disputes such **Decision** before complying with it, the arbitrator may in his discretion award the **Contractor** a reasonable sum in respect of such loss or additional expense, notwithstanding that the overruling of the **Decision** would not otherwise entitle the **Contractor** to any additional payment under the **Contract**.

**49.6** Neither the **Purchaser** nor the **Contractor** shall be entitled to suspend performance of the **Contract** by reason of the reference of a dispute to arbitration.

# Guidance on compiling the Agreement, its Annex, the Specification and Schedules

This guidance has been drafted in the light of English Law in force at the time of publication and in accordance with the principles set out in the Introductory Notes at the front of this publication. The purpose of this section is to help the user to prepare a complete set of documentation to accompany the General Conditions of Contract.

***This guidance is not intended to and cannot give a comprehensive or definitive statement of the commercial or legal position, nor does it form part of the General Conditions or of any contract.***

## The Agreement

In addition to entering the date of the Agreement (usually entered by hand when the document is finally signed), the names and addresses of the parties and the name and location of the Plant, the following must be entered in the body of the Agreement:

- In Paragraph 3: the Contract Price, which if more than one currency is to be used must be stated as the total amount in each of the stated currencies (see also the guidance on compiling Schedule 19 (Contract Price and payment));
- In Paragraph 4: the total aggregate liability of the Contractor to the Purchaser (again this may be stated in more than one currency);
- In Paragraph 5: at (a) and (b) the two sums (again these may be in more than one currency) applicable to the subclauses which are referred to;
- In Paragraph 6: The amount of the Deductible and the applicable currency;
- In Paragraph 8: the date of the Contractor's tender;
- In Paragraph 9: the date for commencement of the Works;
- In Paragraph 10: the name of the Project Manager;
- In Paragraph 11: the name of the Contract Manager;
- In Paragraph 12: the name of the bank (or banks, if there are to be more than one currency of payment) whose base rate will determine the Agreed Rate;
- In Paragraph 13: the amount (expressed as a percentage) of Profit to be added to any Cost – a value which will normally be part of the Contractor's tender;
- In Paragraph 14: the names and/or positions of appropriate senior personnel, their addresses and fax numbers for the receipt of formal Notices – see the guidance below; and
- In Paragraph 16: the seat of any arbitration.

The date for the commencement of the Works is needed in Paragraph 9 because it could be earlier or later than the date of the Agreement, and is relevant to the periods for completion in Schedule 11 (Times of completion). It may also be required for any Special Condition which is included in the Contract.

The requirement to give the name or position, address and fax number of senior personnel of each of the Purchaser and Contractor is to facilitate the issuing of formal Notices in accordance with Sub-clause 1.8 of the General Conditions. While day-to-day communications under the Contract are to be between the Project Manager and the Contract Manager (see Sub-clause 11.8 of the General Conditions), it is recognised that certain communications should be directly between the parties at a senior level – e.g. CEO, Director or Company Secretary. It may be appropriate to state the position of the person (e.g. CEO) rather than his/her name. There are many references in the General Conditions to one party notifying the other (e.g. Sub-clauses 4.3, 7.3, 8.4, 14.5 or 41.7 to list but a few), but these all relate to general matters under the Contract. Notices under Sub-clause 1.8 relate to termination and similar high-level matters which are to be dealt with in a formal way, probably after the taking of legal advice by the party issuing the Notice. Notices issued in accordance with Sub-clause 1.8 are referred to in Sub-clauses 1.1, 14.9 and 41.9, and Clauses 42 (Suspension of the Works), 43 (Termination by the Purchaser for convenience), 44 (Termination for Contractor default), 46 (Disputes), 47 (Adjudication), 48 (Reference to an Expert) and 49 (Arbitration).

If the parties wish the law of the Contract to be other than English law then paragraph 15 should be amended accordingly.

It is advisable that persons entering into a contract ensure that the persons signing on behalf of the parties have the necessary authority to sign if necessary confirmed by documentary evidence.

A further matter which may be appropriate and which should be dealt with in the Agreement is that of conditions precedent. In some instances the parties will need to enter into the Contract by signing the Agreement but then have a period of time within which certain actions must be taken by one or other of the parties, and only when those actions are complete will the Contract come into full effect. Such matters are called conditions precedent. The following clause covers the more frequently required conditions precedent for contracts to deliver major projects. The clause may need adapting to suit the particular requirements. The Parties may wish to consider who will pay for wasted costs if the Contract never comes into full effect. Note the further details that would be required to be included in the Contract according to the clause as drafted.

---

**EXAMPLE ADDITIONAL PARAGRAPH TO INSERT IN THE AGREEMENT**

17. The **Contract** shall come into force on whichever is the later of the following:

    (a) The date specified in the **Agreement** as the date for commencement of the **Works**;

    (b) The date on which any necessary legal financial or administrative requirements specified in the **Special Conditions** as conditions precedent to the coming into force of the **Contract** have been fulfilled.

    The party responsible for meeting any precondition of effectiveness shall notify the other party when the precondition has been met.

    Provided that if all the requirements stated above have not been met and notified by *[date]* this **Agreement** shall be of no effect.

---

## The Annex to the Agreement

The final sentence of Paragraph 1 of the Agreement, following the list of documents which form the Contract, states:

> 'For the purpose of identification, the contents of the **Contract**, including the number of pages in each part, are listed in the Annex to this **Agreement** attached hereto.'

Clarity as to what documents form part of the Contract is essential, as is knowledge as to the extent of those documents – for example, has a page been lost?

The Annex to the Agreement is therefore intended to be a simple list of the various documents which form part of the Contract, with the total number of pages in each document stated. A unique reference number on each page – the Purchaser's contract number for instance – or name of the project or contract will also be helpful in giving clarity as to the identity of those documents.

The Annex itself should be clearly identified as such, and be attached to the Agreement.

## The Specification

*See Sub-clauses 1.1, 1.2, 3.4, 7.1, 8.10 and 33.7*

Every process plant contract needs a Specification, the detailed technical definition of the Plant that the Contractor has agreed to design and construct. The more precise and detailed the Specification of the Plant or any part of it can be made, the greater will be the Contractor's understanding of his commitment and also the Purchaser's definition of his requirements. A comprehensive Specification will enable the Contractor's bid to be fully and firmly based and will provide the criteria against which the adequacy of the finished Plant and its performance, including any Variations, may be evaluated.

Note that Sub-clause 3.4 of the General Conditions states:

> 'The **Plant** as completed by the **Contractor** shall be in every respect fit for the purpose for which it is intended as defined in the **Specification** or in any other provision of the **Contract**.'

A fixed price lump sum contract without an adequate Specification is almost certain to lead to delays and disputes and in all probability higher costs.

The Specification should include all the design standards and codes of practice to which the Plant is to be built including any applicable codes and standards of the Purchaser. It should cover such matters as corrosion allowances and the design life of specific items where applicable. When a plant is to be constructed in accordance with guidelines and codes of practice of national or international authorities, this should be clearly stated in the Specification. Where national standards or codes of practice are issued by authorities such as ASME, BSI or DIN, these are usually sufficiently well-defined not to require interpretation and therefore can be referred to in the Specification with no explanation. However, every contract should be considered individually. If any such authorities may not be recognised by those involved with the Contract, an explanation should be provided.

A Purchaser who is subject to European Procurement Directives, and the national law implementing them, is reminded of their provisions relating to specifications and standards.

In cases where codes of practice, guidelines or standards are less well defined or where Guides to Good Manufacturing Practice are referenced, the Contractor and the Purchaser must take great care to provide sufficient detail in the Specification to avoid any ambiguity in the definition of the standards to which the Plant is to be constructed. In particular, where third party inspectorates will eventually rule on the sufficiency and/or suitability for purpose of the Plant, then this must be clearly stated in the Specification.

Clause 3.2 sets the general requirement on the Contractor to carry out his work 'in accordance with good engineering practice'. This is a common requirement in engineering contracts. However it is not a term that lends itself to a precise, general legal definition.

'Good engineering practice' will vary depending on circumstances. Aspects of good practice in, for example, pharmaceuticals may be wholly inappropriate in oil refining. Should it be necessary to test the concept in any disagreement or dispute, expert evidence will be needed to set the parameters in the appropriate context.

The Specification should be a document which is complete and internally consistent and which:

(a) incorporates all technical details as finally agreed;
(b) identifies clearly and unambiguously the engineering standards that are to be achieved;
(c) defines the battery limits; and
(d) identifies Sections, if any, of the Plant which the Purchaser requires to be completed, taken over or tested for performance as separate Sections.

If early availability of one or more Sections independently of the others is of special importance to the Purchaser, such Sections should be listed in a Special Condition and clearly identified.

The capability of the Plant to make the desired product or products from specified feedstocks will form a principal feature of the Specification, while the required hourly, daily or longer period capacity should be demonstrated in a performance test. It should be recognised that some processes can involve an 'ageing' factor such as catalyst activity. The Specification will need to state an economic working life for any components or materials which have to be replaced periodically. Similarly there may be components of equipment or machinery which have been designed with a limited life, for example filter media or ion exchange resins. These should be listed in Schedule 10 (Parts with limited working life and spare parts) and reference made to the Schedule in the Specification.

The Purchaser may also wish to specify the process performance at the end of the working life of the material or catalyst that requires the longest interruption to production to replace. However it is unlikely that this could be demonstrated by a performance test, because of the time factor. In any case the conditions under which the performance tests are to be carried out, if in any way exceptional, must be stated in Schedule 16 (Performance tests and procedures) (see Sub-clause 35.8 and 35.10).

Where the work involves the modification or extension of an existing plant, the Specification will need to take particular account of the work to be done to that plant. This creates a range of additional considerations, in preparing the Specification, for example:

- battery limits;
- the condition of the original plant, which is in effect being supplied by the Purchaser;
- the presence or suspected presence of hazardous materials (e.g. asbestos);

- any need to maintain production; and
- the interface between the original plant and the new equipment.

Such matters will also need to be covered in the Schedules. Unless the Purchaser is in a position to handover the existing plant to the Contractor for the duration of the work, the Schedules will need to cover the impact on the work of the Purchaser's presence, for example the need to maintain production, the timing of shutdowns and tie-ins, access restrictions and operational hazards.

Consideration should also be given to the question of whether any of the General Conditions, such as Clause 30 (Care of the Works) need to be qualified by a Special Condition. With the inevitable links to insurance (see Clause 31 (Insurance)) appropriate advice should be sought.

## Schedule 1: Description of the Works

*See Clauses 3 (Contractor's responsibilities), 6 (Sufficiency of Contract Price), 7 (Statutory and other obligations), 16 (Variations), 23 (The Site), 26 (Health, safety, environment and pollution), 38 (Final Certificate), 43 (Termination by the Purchaser for convenience) and 44 (Termination for Contractor's default), Guidance on compiling Schedule 3 (Responsibilities of Purchaser) and Guide Note E (Site services and working conditions)*

Whilst the Specification defines the Plant, Schedule 1 should specify the services to be provided by the Contractor including design work, the procurement and construction of the Plant and any temporary works, and any other work to be carried out by the Contractor.

The guiding principle in drawing up this Schedule and Schedule 3 is that they should make clear the dividing lines between the respective responsibilities and scope of work of the Contractor and Purchaser, thereby defining the limits of the Contractor's responsibilities under the Contract.

Many activities concerning the provision of information and designs, the supply of materials and of construction services enter into the execution of any process plant project. They are outlined in the check-list at the end of this guide note. This list is only given as a guide. It is not intended to be complete, nor are the groupings of headings and other details always applicable.

Note that every applicable item must fall within the responsibility of either the Purchaser or the Contractor. There are only two parties to the Contract, the Purchaser and the Contractor. Whatever arrangements either of these parties may have with others, resulting in the provision of certain services by them, those arrangements are subject to separate agreements and form no part of the Contract.

Schedules 1 and 3 should never attempt to give a full list of the contractual activities of both parties, as something may be missed. Any statement in Schedule 1 should be on the lines of 'The Contractor shall undertake all the design, procurement and provision of Materials, construction, erection and testing and other work necessary to complete the Plant and other required work other than that to be undertaken by the Purchaser as detailed in Schedule 3.'

Where the Contractor only supplies a limited service, it would be appropriate to specify his responsibilities in detail in this Schedule. However, with lump sum process plant contracts it is generally the Contractor that is responsible for the majority of the functions. In that case, Schedule 1 could start with:

'Based on the information provided by the Purchaser described in Schedule 3, the Contractor shall ascertain all local conditions relevant to the Works and shall be completely responsible for the design, supply and erection of the Plant including any civil engineering work within the battery limits. The Contractor shall do and provide whatever is necessary to fulfil his obligations under the Contract. He shall also be responsible for all process licences and similar agreements, thereby assuming responsibility for the process design and performance guarantees'.

Schedule 3 should specify the Purchaser's responsibilities in detail.

The approach above would be appropriate for a project on a previously undeveloped site. If a new plant has to be built within an existing facility, the Purchaser may carry out certain other functions such as the provision of underground piping and site levelling. In such a case Schedule 3 should state that the Purchaser will do these things and provide the Contractor with a level site and with the necessary information on ground conditions, on which the Contractor will base his civil engineering design. On some old and badly documented sites, in order to avoid disputes under Sub-clause 6.3 it may be appropriate for the Purchaser to take responsibility for the costs of all groundworks on a cost reimbursable basis. This will require an appropriate Special Condition. Alternatively, the Purchaser may be better served by entering into a contract with a specialist ground clearance contractor in advance of the process contract.

There are occasions when the Purchaser may wish to have the civil work designed and carried out by somebody else. In that case the Contractor may be asked to supply the outline foundation drawings with loads and bolting diagrams. This should be made clear in the text of Schedule 1. Alternatively, the Contractor may be asked to supply the detailed design of foundations for use by the Purchaser or somebody else to carry out the actual civil work.

The process design may be provided by a licensor under a direct contractual arrangement with the Purchaser. For the purpose of the Contract the process information is then the responsibility of the Purchaser unless the Contractor is required in Schedule 1 to verify any design or other information provided by the Purchaser – see Sub-clause 4.7. If detailed process engineering is to be provided by the Contractor it is necessary to define exactly what this comprises in terms of equipment data, detailed flow diagrams, instrumentation details and so on. Schedule 3 should set out the full and precise extent of the process information to be provided by the Purchaser. The Contractor would then do whatever else is necessary, including detailed design, based on this information. Sometimes, however, only the basic process design is supplied by the Purchaser, perhaps from his own research data; under these circumstances the Contractor will have to do whatever else is necessary, but he will start at an earlier point in the sequence of events leading to the final design. The exact starting points for the Contractor should be stated unambiguously, as should the Contractor's responsibility, total or partial, for process guarantees. Great care is needed in writing these critical definitions. If the Contractor offers a Plant complete with process design then he takes full responsibility for this aspect even though the relevant information comes to him from a separate licensor (see also Guide Note A (Innovative work or checks by Purchaser, process licensor or Contractor) and Guide Note B (Intellectual property, know how, confidentiality, information and software)).

When technology is licensed from a third party, whether by the Purchaser directly or by the Contractor as part of his obligations, particular care must be taken to define in the relevant schedule the role of licensor's personnel in checking design and supervising construction, advising on and/or supervising start up and in advising on or supervising any performance tests.

Under certain circumstances the Purchaser may wish to carry out some or all of the construction work himself or else contract for it separately with another party, the Contractor perhaps only being required to provide technical supervision. In such cases Schedule 1 should clearly define the extent of the Contractor's responsibilities and services.

Whilst it is normal for purchasers to take the lead in commissioning, sometimes the Contractor is required to provide additional services particularly in the form of attendance after Plant start-up and during the early months of its operation, while recognising the need of the Purchaser's personnel to have had sufficient training pre-Take Over to enable the Purchaser to fulfil his liabilities upon Take Over (see Sub-clause 33.7); for instance, he may have to carry out an extensive operator and maintenance staff training programme. In such cases the arrangements and responsibilities involved should be defined and detailed in Schedule 9 (Training by Contractor) which should include procedures, key staff to be trained, and duration and timing of the training. This training requirement should extend beyond plant operation to include health and safety, quality assurance and maintenance procedures.

If it is considered by either party that it is not possible to define clearly what has to be carried out, it may be appropriate to define broadly what is required of the Contractor and cover that requirement by a provisional sum, in which case a Special Condition should be included – see guidance on compiling Schedule 19 (Contract Price and payment).

'Qualification procedures' and 'validation procedures' are becoming increasingly important, particularly when the Purchaser requires the Contractor to achieve a specific standard of quality either defined in a national or international standard or when a third party regulating body insists on prescribed and highly-developed testing procedures to ensure that the Plant operates satisfactorily within the tolerances stated in the Specification. Qualification and validation apply to all phases of the development of a plant, including design, installation, operation and process. It is normal for the Contractor to be required to carry out the qualification up to and including operation of the plant but usually excluding qualification of the process, particularly if the process or technology is owned by the Purchaser. Schedule 6 (Quality assurance and validation) should outline the Purchaser's requirements.

Schedule 1 should include a description of the Site (see Sub-clause 1.1 definition of 'Site'). The description should include:

(a) a written description which covers location, map references, what the Site comprises;
(b) plans of the Site including maps and layout, clearly indicating any hazardous, environmentally sensitive or 'no go' areas;
(c) any weather or seismic criteria which could affect the Works;
(d) information on current and former uses, buildings, buried structures.

Where the Site is an existing plant, it is clearly important for Schedule 1 to provide detailed information concerning the topography and characteristics of the Site, including all available information on former uses, buildings and facilities located on it.

The details about site access necessary in Schedules 1 and 3 will vary depending on whether the Site is undeveloped or part of an existing facility. In Sub-clause 23.1 it is assumed that the Site belongs to the Purchaser and that if necessary the Purchaser will arrange that the Site includes a suitable access from the public highway.

In this context access could mean a clear access route over land without any road paving or railway track, in which case Schedule 1 could include the construction of an access way as part of site preparation. Alternatively the Purchaser could have arranged for a third party to construct the access way ready for use by the Contractor. Either way, such an access way should form part of the Site. In any case it will be essential for the Contractor to ensure that the access way has the capacity required, both in terms of weight and clearances for large loads on transport, for the passage of the largest loads to be brought on to the Site. The Contractor will have to take account of any limitations to access, such as the maximum size of load that can be transported around the Site, for example, by allowing for a greater amount of site assembly than he would ideally like. While it is the Purchaser's duty to provide a suitable right of access, just what is suitable will be for definition before award of the Contract having regard to the equipment, materials and labour to be brought onto the Site. Contractors should also be aware that if large loads are to be brought to site from a port, railhead or public road system, it is essential to investigate and if necessary incorporate into the design and planning any limitations that may be imposed by size, weight or movement control rules.

If the site is part of or immediately adjacent to an existing installation, there may be specific constraints on access arising from health and safety or operational issues.

All documentation necessary to complement the information contained in the Schedule should be listed and attached to the Schedule.

**Check List**

This list of activities and outputs is given as a guide only as to the range of activities and outputs which may be included in Schedule 1. Generally those at (1) Preliminary activities and (2) Definition of project data would need to be carried out prior to inviting tenders.

(1) Preliminary activities, such as:

- (a) local surveys and site investigations;
- (b) site conditions – climate, access, ground condition, labour situation;
- (c) contaminated ground and presence of hazardous materials.

See also (15) Other activities

(2) Definition of project data:

- (a) process requirements;
- (b) raw materials, feedstocks, utilities;
- (c) products, intermediate products and by-products;
- (d) effluents, with limitations on discharges and emissions (see Schedule 5 (Environmental protection & waste disposal));
- (e) battery limits (see the Specification);
- (f) battery limit conditions for (c), (d) and (e) above;
- (g) statutory requirements.

(3) Safety (see Schedule 4 (Health & Safety)):

- (a) safety review records;
- (b) risk assessment/safety case;
- (c) assessment of hazardous substances including any requirements of applicable legislation;
- (d) chemical data sheets;

(e) Hazard Analysis Assessment (HAZAN) and Hazard Operability Study (HAZOP);
(f) safety review procedure (including major hazard site assessment);
(g) disaster plan;
(h) safety plan and risk assessment including any requirements of applicable legislation.

(4) Process design:

(a) basic flowsheets;
(b) preliminary process calculations;
(c) heat and material balances;
(d) hazardous materials data sheets;
(e) process design manual (see Schedule 2 (Documentation)).

(5) Detailed chemical engineering design:

(a) detailed process calculations;
(b) equipment data;
(c) process flow diagrams;
(d) piping and instrumentation diagrams;
(e) instrumentation and control data;
(f) process control computer interfacing;
(g) materials of construction.

(6) Detailed mechanical and structural engineering design (see also (8) below):

(a) equipment list;
(b) equipment sketches or drawings;
(c) equipment data sheets;
(d) general arrangement and detailed layout;
(e) piping arrangements and details;
(f) models;
(g) line diagrams (piping, mechanical, etc.);
(h) structural steelwork;
(i) insulation and painting requirements;
(j) engineering standards;
(k) labelling and numbering system;
(l) heating, ventilation and air conditioning system design;
(m) room data sheets including environmental standards.

(7) Electrical installation design requirements:

(a) single line diagrams;
(b) hazardous area electrical classification;
(c) list of electric motor drives;
(d) power distribution scheme;
(e) incoming power supplies;
(f) electrical sub-stations and transformers;
(g) sub-distribution boards;
(h) uninterruptible power supplies;
(i) emergency power supply scheme;
(j) emergency generators;
(k) wiring and cable standards;
(l) motor control centres;
(m) labelling and numbering system (see also 6(k) above).

(8) Control and instrumentation:

(a) control and instrumentation philosophy;
(b) single line diagrams;
(c) control loop diagrams;
(d) computer hardware and software standards;
(e) control logic diagrams;
(f) instrument lists;
(g) data collection and recording requirements;
(h) fire detection system;
(i) security and personnel access system;
(j) telephones and communications;
(k) gas detection system;
(l) operator/control interface;
(m) discharge monitoring and alarm systems for hazardous or statutorily limited discharges.

(9) Civil engineering and building design:

(a) outline foundation and building drawings;
(b) detailed design of foundations, structures, buildings;
(c) roads, trenches, earthworks;
(d) specifications, bills of quantities;
(e) documentation for planning approval;
(f) documentation and calculations for building regulations approval;
(g) drainage (domestic, storm, foul and process);
(h) effluent holding and treatment system.

(10) Project co-ordination procedures (see Schedule 20 (Contract co-ordination) and Schedule 21 (Reports and records)):

(a) project programme;
(b) reporting;
(c) cost report;
(d) procurement;
(e) communication, documentation, distribution and transmittal procedures (see Sub-clause 11.8 and Guidance on Schedule 20 (Contract co-ordination));
(f) document identification and numbering system;
(g) quality assurance standards to be adopted (see Schedule 6 (Quality assurance and validation));
(h) validation.

(11) Supply of plant, equipment and other materials, packaged units, chemicals and catalysts (see Schedule 7 (Subcontracting)):

(a) enquiring and purchasing;
(b) expediting;
(c) inspections—statutory and non-statutory;
(d) factory and site acceptance testing (see Schedule 13 (Pre-installation tests and procedures));
(e) shipping services, transport to Site;
(f) test certification;
(g) approved vendors lists;
(h) special tools and spare parts;
(i) chemicals and catalysts.

(12) Construction aspects (see Guide Note E):

(a) access to Site;
(b) off-loading, reception and storage of materials;
(c) construction manpower histograms identifying trade disciplines and training/certification requirements;
(d) Contractor's Equipment and tools;
(e) supervision at various levels (see Schedule 8 (Contractor's named personnel));

(f) services and utilities;
(g) subcontracts;
(h) civil engineering works;
(i) Site administration including normal working hours (see Sub-clause 28.5);
(j) Site conditions;
(k) Site personnel policy;
(l) Site safety policy (note that the Purchaser's responsibility for providing copies of all his rules affecting access, safety, working methods or industrial relations should be stated in detail in Schedule 3);
(m) precautions to be taken regarding contaminated soil, adjacent premises, etc.

(13) Preparation for start up (see Guide Note J (Completion, Taking Over, testing and start up)):

(a) operating manuals;
(b) maintenance manuals;
(c) mechanical testing (including pressure testing, testing of lifting beams and other equipment, etc.) (see Schedules 14 (Criteria for the completion of construction) and 15 (Take Over procedures));
(d) Plant operators;
(e) supervision;
(f) specialists;
(g) training for all personnel (see Schedule 9 (Training by Contractor));
(h) process materials, lubricants;
(i) spares availability and store keeping (see Schedule 10 (Parts with limited working life and spare parts));
(j) safety system tests and records.

(14) Start up, Plant operations:

(a) operating labour;
(b) supervision;
(c) specialists;
(d) laboratory and analytical services;
(e) routine maintenance;
(f) alterations, construction services;
(g) performance tests (see Schedule 16 (Performance tests and procedures));
(h) Plant performance studies (see Schedule 17 (Performance guarantees and damages for failure));
(i) emergency procedures;
(j) Plant shutdown procedures.

(15) Other activities:

(a) environmental impact analysis/assessment;
(b) insurance policies (see Clause 31 (Insurance));
(c) statutory inspections and licences;
(d) statutory application and permit to discharge effluent;
(e) control and recording of waste disposal (see Schedule 5 (Environmental protection & waste disposal));
(f) planning and other permissions and consents including permits required for use of the Site (see Sub-clause 7.1).

Schedule 1 is referred to in Clause 1 and Sub-clauses 7.1 and 28.5, and more generally the Works are referred to in numerous places throughout the General Conditions.

# Schedule 2: Documentation

*See Clauses 20 (Confidentiality) and 21 (Documentation)*

### Documentation for approval

The Project Manager will almost certainly wish to approve certain key drawings or other documents. The procedure for submitting and approving them is laid down in Clause 21 but the actual documents which are to be subject to this procedure should be listed in this Schedule under the heading 'Documentation for Approval'. While recommendations can be made that are not always applicable, the following should be considered for inclusion in the list:

(a) drawings or other documentation required in support of submissions to local and other statutory authorities;
(b) process diagrams that may typically include material and energy balances, product quality, critical flow characteristics and battery limit or interface conditions;
(c) process control and instrumentation diagrams;
(d) layouts (plot plans and elevations);
(e) drawings locating battery limit crossings;
(f) electrical single line diagrams;
(g) safety plan.

Important design calculations, specific design details of critical equipment items or other technical documents may also require the Project Manager's approval.

**Whenever possible the essential Documentation should be discussed and agreed during the tender negotiations and incorporated into the Specification, so that it becomes part of the Contract and needs no further approval. The Documentation to be approved later should be kept to the minimum in order to avoid hindering the progress of the project.**

### Documentation for information only

If the Project Manager wishes to see certain types of Documentation under the terms of Sub-clause 21.7 as a matter of course, these may be listed in this Schedule under the above heading, thus avoiding the need for the Project Manager to give notice on each occasion. The list should include the type of Document, the subject to which it refers, the stage of its completion and the form in which it should be submitted.

### Final Documentation and Manuals

These are referred to in Sub-clause 21.9 and are likely to include drawings necessary for the Purchaser's operation and maintenance of the Plant, mechanical manuals, spares lists, start up and operating instructions and similar documents. It is now commonplace for the Documentation to be supplied under this Sub-clause in electronic format in addition to agreed numbers of printed versions. It is therefore important that details such as the number of copies, type of reproducible, programme, etc. should be specified in this Schedule under the heading 'Final Documentation and Manuals'. 'As built' drawings will normally be required. If the Purchaser does not require these to be provided by the Contractor it should be expressly stated.

The Purchaser should review, after perhaps six months of commercial operation, how the Plant is actually being operated and whether or not it complies with the designer's intention as regards both safety and performance, and whether or not the Final Documentation and Manuals need modification in the light of practice. It is also useful for the Contractor to revisit the Plant periodically to gain and collate practical feedback on Plant operation, for the benefit of future design (subject to any constraints of commercial confidentiality).

The operating manual will describe how the process is designed to work, how to start up the Plant and shut it down, and what should be monitored while it is running to achieve the maximum available efficiency. The operating manual must also contain specific instructions and information concerning emergency shut down and start up, particularly where hazardous materials are being handled. It is also vital to the safe operation of the plant that the Purchaser, who is normally responsible for operating activities, is fully aware of the detailed parameters and operating principles which are the basis of the design.

If, after the Plant has been running for some time, a change takes place in some external factor such as feedstock composition, it may be necessary to review the process design in order to determine new operating conditions which will accommodate the change and still optimise Plant performance. Alternatively, deterioration in performance may occur without any obvious cause, or, again, the Purchaser may call for a review of capacity. The resulting 'troubleshooting' and 'debottlenecking' exercises will require the Purchaser to know more about the original Specification than simply the final equipment data; he will need to know what compromises were made between minimum capital cost and minimum operating cost, what safety margins were used in arriving at sizes or thicknesses of materials subject to stress or corrosion, what additional heat transfer surface became available by virtue of the use of standard tubing larger than the minimum theoretical size, and so on.

It is therefore recommended that in addition to the operating, maintenance and safety manuals, a process design manual should be included in the list of Final Documentation and Manuals. The process design manual should include the background information on process design, and particularly describe how compromise decisions have been made. A similar consideration applies to computer software. In the event that changes are made to the plant design or to its as-built state, these changes should be clearly recorded and then included in the final issue of the process design manual.

**Documentation not to be shown to the Project Manager**

Note that Sub-clause 21.6 excludes certain categories of document from the Project Manager's rights of inspection. If any documents are to be excluded they should be described in this Schedule or in a Special Condition.

Schedule 2 is referred to in Sub-clauses 21.2, 21.6, 21.7 and 21.9.

## Schedule 3: Responsibilities of Purchaser

*See Clause 4 (Purchaser's responsibilities) and Guidance on compiling Schedule 1 (Description of the Works)*

It is rare for the Purchaser to provide no input into the design, engineering and construction aspects of a process plant. This Schedule should list the activities, information and supplies for which the Purchaser will be responsible, and the times at which these should be completed so that the Contractor's progress is not delayed. It is important that the Purchaser appreciates that if he has an obligation to provide specific items, particularly information, for the Contractor, timely performance of this obligation is essential. It should be noted that although Schedule 1 may contain outline information about any design and engineering to be carried out by or on behalf of the Purchaser, the details must be set out in this Schedule.

The Purchaser must also be aware of his obligation under Sub-clause 7.1 to obtain all permits in connection with the use of Site for construction, operation and maintenance of the Plant except for those expressly stated in the Contract to be the responsibility of the Contractor. It is recommended that all the permits which the Purchaser has or will obtain are listed in Schedule 3, complete with an indication:

(a) that they have been obtained and details of any conditions stated in any permit, or
(b) that they will be obtained and details of when the Purchaser expects to obtain them.

The latter may link to the Purchaser's obligations under Sub-clause 4.1 or 4.2(c).

In most cases the Purchaser will provide information in advance of the Contractor starting work. Some of this may be contained in the Specification and its attachments, but all the Documentation to be provided after the date of the Agreement should be described in as much detail as possible in this Schedule.

The list of activities provided at the end of Schedule 1 is provided as an aide-memoire in checking that all the relevant activities to be undertaken by the Purchaser (other than those specifically mentioned in the Contract Conditions) are included in this Schedule.

If technology is licensed by the Purchaser from a third party, this Schedule should define the role of the licensor and what documentation is to be supplied as part of any licensing agreement. In addition, the role of the licensor's personnel in checking or advising on design work, specifications, construction, start up and performance testing must be set out in detail.

In some instances, the Purchaser may elect to procure certain items of equipment and 'free-issue' them to the Contractor. The 'free-issue' items may, for example, be special process equipment, which the Purchaser wishes the Contractor to install, or 'long lead' equipment items that have to be ordered by the Purchaser before he has engaged the services of the Contractor. The items must be clearly identified and delivery dates specified. Occasionally, the Purchaser may not wish to disclose details of a specific unit process or item of equipment for reasons of confidentiality. Under such circumstances, sufficient information concerning interface data must be supplied to allow the Contractor to provide the necessary infrastructure to support the operation of the items supplied by the Purchaser.

Consideration was given to the inclusion in the General Conditions of a defined term 'Free-Issue Materials', but the view was taken that such a term is not generally necessary at main contract level (such a term is included in the Yellow and Brown Books Forms of Subcontract). However, if there are to be a number of different 'free-issue' items, it may be beneficial to have such a defined term for such items which could then be specifically listed in Schedule 3. A suitable definition, to be included in Sub-clause 1.1, would be:

> '**Free-Issue Materials**' means machinery, plant and other items of equipment and materials intended to form part of the **Plant** and anything needed for their use, to be supplied by the **Purchaser**.

It should be noted that the definition does not cover any item of equipment which the Purchaser might make available to the Contractor to use in constructing or erecting the Plant, i.e. equipment to be used in place of Contractor's Equipment. If the Contractor is to be allowed use of such equipment (e.g. an overhead gantry crane within an existing building) then that should also be detailed in Schedule 3, but it would not constitute Free-Issue Materials as it would not form part of the Plant.

In determining whether or not to use the expression '**Free-Issue Materials**', consideration should be given to the adequacy of the following clauses in defining the supply of, ownership of and responsibility for such 'free-issue' items:

Clause 4 (Purchaser's responsibilities)
Clause 24 (Delivery to Site)
Clause 25 (Ownership of Materials)
Clause 30 (Care of the Works)
Clause 31 (Insurance)

In Clause 23 (The Site) it is stated that the Purchaser will give possession of the Site to the Contractor, and access to it. This may involve the construction of a roadway, gates, fencing etc. The extent to which these will be provided by the Purchaser should be specified in this Schedule.

Where the Site is part of an existing facility of the Purchaser, with access controlled by his security staff, the Purchaser will need to issue instructions regarding the availability of permits for individuals and vehicles to enter or leave the premises, and also about observing any restrictions on access to other parts of the facility. The latter should be clearly defined by suitable signs.

Copies of any rules applying to personnel on the Site must be supplied by the Purchaser to the Contractor. If compliance with any of these could increase the Contractor's costs, they should form part of the enquiry against which the Contractor is invited to quote. Such rules could affect safety, working methods or industrial relations.

The Purchaser may choose to provide security, welfare and other facilities for use by the Contractor. If so, these should be detailed in this Schedule.

The extent of the Purchaser's responsibilities for Health and Safety (see Guidance on compiling Schedule 4 (Health & Safety)) should be stated in this Schedule, as should the extent to which the Purchaser will be responsible for the disposal of products and solid or liquid wastes arising from the operation of the Plant, and the circumstances under which this will apply (see Guidance on compiling Schedule 5 (Environmental protection & waste disposal)).

Great care must be taken to ensure that the respective responsibilities of the Contractor and Purchaser for checking designs for quality and completeness of their work are accurately and fully defined.

Schedule 3 is referred to in Sub-clauses 4.1, 4.2, 24.1 and 27.4.

## Schedule 4: Health & Safety

*See Sub-clause 7.2 and Clause 26 (Health, safety, environment and pollution)*

In the United Kingdom, the Health and Safety at Work, etc. Act 1974 (as amended by Schedule 3 to the Consumer Protection Act 1987) imposes obligations on the Purchaser, the Contractor, the Subcontractors and others to ensure that the Plant is safe to operate and maintain and that adequate documentation and other information is provided to ensure that the Plant can be operated and maintained safely and without risks to health. It is not possible to transfer either the obligations or the penalties for failure to meet them. In the case of a process plant, however, there is room for genuinely held differences of opinion as to what is or is not 'reasonably practicable' or 'safe' or 'without risk to health'. It is essential that neither the Purchaser nor the Project Manager can impose contractually upon the Contractor a requirement to do something that the Contractor considers to be a risk to health or safety.

In drawing up the Specification and Schedule 1 (Description of the Works), both parties should satisfy themselves that the health and safety risks are fully assessed and eliminated so far as reasonably practicable. The parties should remember that in the event of any incident both the Purchaser and the Contractor may be liable to prosecution under the Act.

It should be noted that Clauses 3 (Contractor's responsibilities), 16 (Variations) and 17 (Variations proposed by Contractor) are particularly relevant where any unforeseen hazard comes to light in mid-project. Risks to health or safety are hazards and should be treated as such. Rather than run any risk of borderline areas being neglected, a degree of overlap of the work of the Contractor and the Purchaser when checking designs and the quality and completeness of work should be encouraged.

All relevant United Kingdom legislation should be taken into account by the parties, for example the Management (Health and Safety) Regulations 1999, and the Construction (Design and Management) Regulations 2007 (CDM Regulations), the Control of Major Accident Hazard Regulations 1999 (COMAH) as amended in 2005. Specific appointments must be made by the Purchaser in order to comply with the CDM Regulations. It is to be expected that the Contractor will act as Principal Contractor under the CDM Regulations up to Take Over. This Schedule should state clearly who is the Principal Contractor.

Schedule 4 is referred to in Sub-clauses 4.6, 7.2, 26.3 and 26.4.

## Schedule 5: Environmental protection & waste disposal

*See Clauses 26 (Health, safety, environment and pollution) and 34 (Site clearance)*

The impact of environmental protection on the design and operation of manufacturing plants has become highly significant. As a result, all parties should be aware of any national and international laws and treaties which affect the project.

For those projects which could have an impact on the environment, it is the duty of both the Purchaser and the Contractor to establish a properly considered analysis of the impact on the environment of the plant, its products, by-products and wastes. Whilst some issues are relatively simple, requiring disclosure of the level of release to the environment or statements regarding volumes, analysis and frequency of release, others are significantly more complex. In the latter case a detailed Environmental Impact Assessment is usually required by environmental protection agencies. This should normally be undertaken by a qualified specialist employed or retained by the Purchaser.

The ultimate responsibility for pollution usually lies with the operator of a site or plant. This is particularly true regarding issues such as the following:

(a) disposal of waste;
(b) contamination of air, sea, land and ground water;
(c) disposal of packaging materials or product containers;
(d) the handling and disposal of nuclear and radioactive chemicals.

Item (d) needs special consideration due to the potentially long period of contamination resulting from the radiation caused by nuclear materials. It should be noted that there is a statutory duty to consult the Nuclear Inspectorate regarding the handling, storage and disposal of radioactive material. Specialist advice should always be sought in the event that either the Purchaser or the Contractor is not sufficiently experienced in the handling and processing of hazardous materials.

In addition, the use of recycled building materials in construction to minimise the use of non-renewable or scarce resources is increasingly required.

Sub-clause 26.7 sets out indemnities between the parties as to which party will bear the cost of any clean-up of pollution. Note that there is nothing about who might do the clean-up (which might be done by a third party), simply an allocation of the cost.

Clause 34 places the clear responsibility for Site clearance with the Contractor. However, it must be recognised that there are a variety of types of waste other than those specifically stated which may require disposal. Typical forms of waste may be as follows:

(a) *Standard non-hazardous waste resulting from the construction of the Plant:*
The management of this waste is generally the responsibility of the Contractor, as detailed in Schedule 5.
(b) *Waste resulting from Site preparation or demolition which may be a mixture of hazardous and non-hazardous materials:*
Asbestos, toxic or biological wastes are typical of the hazardous materials which may be found on an existing chemical manufacturing site. The Purchaser normally bears the responsibility of ensuring that such hazardous materials are removed from Site safely and disposed of at a designated or licensed site. The Purchaser may request the Contractor to organise the necessary specialists to carry out this work and it is expected that in so doing he will operate all necessary safety procedures to comply with legislation and the Codes of Practice which apply. Nevertheless the ultimate responsibility still lies with the Purchaser and he must oversee each stage of the demolition or Site clearance where hazardous materials are being handled, to satisfy himself that he is discharging his legal obligations in this regard.
(c) *Substances which are hazardous, toxic or possibly of unknown character resulting from operation of the Plant:*
In this instance it is normally the sole responsibility of the Purchaser to arrange for the safe removal and disposal of waste. An exception to these circumstances would be if the Contractor is the owner of the process technology, in which case he may be best placed to establish the correct protocols for the removal from Site of any waste, by-product or generally hazardous materials from his process, but the actual removal and disposal of these will be the responsibility of the Purchaser from the time he has taken over the Plant.

Schedule 5 is referred to in Sub-clauses 4.6, 7.2 and 26.4.

## Schedule 6: Quality assurance and validation

*See Clauses 22 (Inspection and pre-installation tests), 32 (Completion of construction) and 33 (Taking Over) and Sub-clause 3.8*

In broad terms, quality assurance requires a documented systematic approach to the attainment of quality.

Quality assurance is now almost universally required in process projects. 'Quality Assurance' can be defined as all activities and functions concerned with the attainment of quality and 'Quality' as the totality of features and characteristics of a product or service that bear on its ability to satisfy a given need.

There are several recognised quality assurance standards. ISO 9001 'Quality Systems: Part I Specification for design/development, production, installation and servicing' is particularly applicable to the process industry.

In addition to the standards for quality, specific engineering or other standards are likely to be listed in the Specification. Care must be taken to define to what extent each party is responsible for producing record documents.

The Purchaser must advise the tenderers at the pre-contract stage of any specific quality assurance issues that are to apply to the project. He should provide copies of quality assurance control procedures that the Contractor will be required to adopt; for example:

> 'The Contractor shall compile indexed and cross-referenced records comprising such Documentation as drawings, non-destructive and intrusive test records, and material and test certification. Such records shall be subject to review and approval by the Project Manager in accordance with Sub-clause 21.2 of the General Conditions.'

Before completing Schedule 6, detailed discussions should take place between the Purchaser and the intended Contractor regarding the extent and complexity of the quality assurance activities which will be required during the project. As a minimum, the contents of a master plan should be included in the Schedule, with a tabulation of the responsibilities of the Purchaser, Contractor, Subcontractors and suppliers clearly allocated. The Schedule should also state whether third party inspectors or quality assurance specialists are to be involved and what responsibilities they will have for approving documentation. If the quality assurance activities are critical for completion, the times and duration of such activities should be included in the Approved Programme.

A quality assurance master plan should also include sections on qualification activities which, when they have been completed, will provide detailed documentation on all the checks that have been carried out during design, procurement, installation and testing to ensure the robustness and reproducibility of the process when in operation. It is emphasised that prior to establishing a master plan, the Specification and Schedule 1 (Description of the Works) should be thoroughly reviewed and documented since these essential documents will form the basis for any quality plan.

When an event such as payment or Taking Over is dependent upon quality assurance (or validation) activities, this should be clearly cross-referenced in the appropriate schedule.

Validation may be required as part of the overall quality assurance programme. The purpose of validation is to create documentary evidence providing assurance that the product from a manufacturing process will consistently meet its defined specifications and quality characteristics in all circumstances. In particular the pharmaceutical, electronics and aerospace industries apply rigorous techniques of this nature. If the Purchaser or the Contractor is unfamiliar with such activities, they should engage expert advice before signing a contract in which Validation is included since the time and effort necessary to establish fully compliant procedures and documentation is often underestimated.

If it is to be applied, validation needs to start at the development phase (i.e. pre-contract) by defining the product specifications in all respects, including the deviations which can be tolerated, but which must never be exceeded when the Plant is in operation. The principal purpose is to ensure that the Plant meets the design intent in every respect.

The following schedules may also refer to quality assurance activities:

Schedule 2 (Documentation);
Schedule 13 (Pre-installation tests and procedures);
Schedule 14 (Criteria for the completion of construction);
Schedule 15 (Take Over procedures);
Schedule 16 (Performance tests and procedures).
Schedule 6 is referred to in Sub-clauses 3.8, 22.2, 22.3, 32.3, 32.5, 33.3 and 33.5.

## Schedule 7: Subcontracting

*See Clauses 9 (Subcontracting and third party rights) and 10 (Nominated Subcontractors)*

In process plant projects most, if not all, of the equipment and other materials will be designed, manufactured and supplied by a number of Subcontractors. Similarly, some or all of the construction work on Site will normally be subcontracted.

Some critical items are often ordered by the Purchaser pre-contract as long-lead items or the Purchaser may authorise the Contractor to order them prior to contract award. If the Purchaser has ordered them such materials are either to be 'free issue' under Sub-clause 4.2 or the order will need to be novated to the Contractor after the Contract is made. Such an action will require a Special Condition. Before agreeing to the novating of a Subcontract previously established by the Purchaser, the Contractor should consider the risks associated with a Subcontract to which he had no pre-contract input.

The Purchaser may wish to ensure that the design of a particular process unit or element of the Plant is the responsibility of designers with appropriate specialist experience. The Purchaser should make this clear at the tender enquiry stage. A contractor who is tendering for a project may himself have some specialised design experience, but may need to subcontract other parts of the design to specialist Subcontractors, in which case he should make this clear in his tender. In this context the Purchaser may require direct access to the Subcontractor to support post completion activities at the Site, to resolve technical problems or to assist in modifying equipment or systems supplied. Under such circumstances the relationship between the Purchaser and Subcontractor should be the subject of a Special Condition incorporated in the appropriate Subcontract documentation. All such detailed requirements should be clearly stated in Schedule 7. (See also Guide Note R (Rights of third parties)).

If the Purchaser has stated in the tender enquiry that he requires certain design work to be carried out by named specialists, they should be nominated by the Purchaser in this Schedule.

If the Contractor has stated in his tender that he proposes to subcontract parts of the Works to named specialists, and the Purchaser accepts them, their names and a brief description of their scope of supply should be added to this Schedule before contract award.

Sub-clause 9.2 makes it clear that the Contractor carries the responsibility of fulfilling the terms of the Contract irrespective of the performance of any Subcontractor. Sub-clause 9.3 requires the Contractor, subject to anything otherwise agreed with the Project Manager, to ensure that each of his Subcontractors is required to observe equivalent contractual obligations to those set out in the General Conditions and in any Special Condition, and to observe any provisions in the Contract which apply to subcontracts. While it may not be possible for the Contractor to ensure this in each and every case, for example with reference to the Defects Liability Period or in the event of termination of the Contractor's employment (for either of which he should obtain the Project Manager's dispensation) this does mean that the Contractor must structure his subcontracts to reflect the obligations of the Contractor. This is most easily achieved by using the IChemE Forms of Subcontract, as described below.

The Contractor's responsibilities in respect of nominated Subcontractors are set out in Clause 10. In particular it specifies the procedure to be adopted if the Contractor has valid reasons for objecting to any Subcontractor nominations which the Project Manager has made. If the Purchaser or the Project Manager nominates a Subcontractor then the resulting Subcontract will be between the Contractor and the Subcontractor, and that all subsequent commercial matters such as Variations and guarantees, payment, and rights of inspection and expediting shall be directly between the Subcontractor and the Contractor so that the latter can fulfil his obligations.

No further definition is required where particular makers of equipment or other Subcontractors are named in the Specification. If the Purchaser wishes further to restrict the Contractor's right to subcontract or to have free choice of Subcontractors, the basis of such restrictions should also be stated in this Schedule. It may be convenient to have lists of approved suppliers for different classes of equipment and materials from which the Contractor is free to choose. Provision may be made for adding names to the list by agreement with the Project Manager.

A Purchaser who is subject to particular laws, such as EU Procurement Directives, is reminded of the need to comply with such directives and regulations, which may include rules concerning the nomination of Subcontractors, for example that Subcontractors or particular makes of materials are not to be named in the Contract.

IChemE publishes a Form of Subcontract consistent with and suitable for use with these conditions (the Yellow Book) and a Form of Subcontract for civil engineering works consistent with and suitable for use with these conditions (the Brown Book).

Schedule 7 is referred to in Sub-clause 9.2.

## Schedule 8: Contractor's named personnel

*See Clause 12 (Contract Manager and Contractor's staff)*

Many Purchasers regard the experience of a particular Contract Manager to be of prime importance. Therefore there is a provision in Sub-clause 12.1 that the Contract Manager will not be replaced without the consent of the Project Manager.

Where specialised experience is required to carry out a project, the Purchaser may consider it desirable or in some cases essential to request the Contractor to nominate not only the Contract Manager but also key engineering and/or construction supervisory staff. Sub-clause 12.4 provides for the continuing participation of named key personnel who are of considerable importance to the success of the project. These nominations should be agreed pre-contract. A list of such named personnel will be included in this Schedule.

Sub-clause 12.4 recognises that the Contractor may wish to replace key personnel identified in the Contract, but such replacement must be with the prior consent of the Project Manager. If necessary, this Schedule should specify the circumstances under which the Contractor may nominate a replacement and any required overlap period for handing over to the replacement. The Project Manager will be expected not to withhold his consent unreasonably. A suitable phrase can be 'which shall not be unreasonably withheld in the event of a personal reason, resignation, or other circumstances outside the Contractor's control.' Possible promotions are a more difficult area, as they will be within the Contractor's control, but equally the Contractor has to be aware of employment law and not withhold a promotion in a discriminating way. In allowing individuals to be changed it is common to have a requirement that the replacement person will be equally or better qualified and experienced than the person who is being replaced. The Purchaser may wish to ensure that the likelihood of the removal of key personnel without his consent is minimised. To this end, the Purchaser may wish to introduce a Special Condition calling for a payment to be made by the Contractor in the event that he removes key personnel without the Project Manager's consent. The amount of the payment could be related to the annual salary of the person changed.

Schedule 8 is referred to in Sub-clauses 12.3 and 12.4.

## Schedule 9: Training by Contractor

*See Sub-clauses 3.9 and 4.5*

Sub-clause 3.9 requires the Contractor to provide training for operating and maintenance personnel, and in return the Purchaser has the responsibility to make suitable personnel available for training. The best arrangement for practical training to be achieved is for the Purchaser to appoint and make available the key operating and maintenance personnel before completion of construction and well before commissioning so that they can work alongside the Contractor's engineering, construction and commissioning personnel. For example, they can assist the construction personnel in making lists of outstanding minor items requiring modification or finishing. This should familiarise them with the equipment, control systems, safety systems, pipework and the location of valves, vents and drains.

Training covers a broad area. The Contractor may be called upon to provide basic and practical training on various aspects of Plant operation, health and safety, quality assurance, instrumentation, control systems and maintenance procedures for various categories of personnel. Where the Plant includes technology with which the Purchaser's personnel are not familiar, induction courses and workshops should be arranged explaining the process, the philosophy of design and control systems, and the start up, shutdown and emergency procedures.

All these requirements should be listed in the Schedule, together with:

(a) the methods and procedures to be adopted in carrying out the training, the personnel who are to be trained and

(b) the timing and duration of the training.

The Contractor in consultation with the Purchaser should set minimum acceptable levels of competency to be achieved at the completion of each type of training and for any testing of personnel.

If for any reason no training is required Schedule 9 should clearly state that no training is required.

Schedule 9 is referred to in Sub-clauses 3.9 and 4.5.

## Schedule 10: Parts with limited working life and spare parts

*See Clause 37 (Liability for Defects) and Guide Note L (Liability for Defects)*

Leaving aside normal wear and tear, it may be appropriate to specify certain replaceable parts such as filter media or ion exchange resins, with a short working life because the cost of a more durable alternative would be many times higher or may be technically inappropriate. Such parts should be detailed in this Schedule. A suitable introduction to a list of such parts could be:

> 'The following parts have been specified recognising they have a limited working life. Attention is drawn to Sub-clause 37.11 of the General Conditions in respect of such parts.'

The list should identify the item of equipment in which the part is fitted, a description of the part including any identification code, the estimated working life of the part and the guaranteed working life under normal operating conditions.

Where the guaranteed working life of a part is less than 365 days, the Defects Liability Period of that part is reduced to that working life in accordance with Sub-clause 37.11.

It is normal for the Purchaser to specify that the Contractor shall provide spares, typically to cover two years of operation, in addition to commissioning spares. This requires extensive discussion and detailed agreement between Purchaser, Contractor, and Subcontractors to establish a basis for deciding on the quantity and type of spares to be provided. Risk and reliability analysis is one method often used.

Spares will vary widely in quantity, size and price. It is, for example, common for major rotating equipment to be sent to Site with spare rotors. Such items need careful warehousing and maintenance, such as regular turning or barring over.

Other types of spares are needed for all rotating equipment, control valves, instrumentation, electrical items, etc. Often the classification and coding of all the many spare parts have to fit into the Purchaser's established inventory system. It is easy to underestimate the cost of this aspect of a project.

In some cases the Purchaser may wish to set up a 'just-in-time' spare parts supply chain. In this case clear delivery instructions, together possibly with detailed storage descriptions and minimum inventory levels need to be agreed.

Schedule 10 is referred to in Sub-clause 37.11.

## Schedule 11: Times of completion

*See Clauses 4 (Purchaser's responsibilities), 13 (Times of completion and Approved Programme), 16 (Variations), 21 (Documentation), 32 (Completion of construction), 35 (Performance tests), Sub-clauses 7.3 and 39.8, and Guide Note J (Completion, Taking Over, testing and start up)*

The Contractor's work may have a single objective of significance to the Purchaser in terms of timeliness of completion, such as the construction and start up of a process unit; alternatively it may have several such objectives of varying priority. These will form the main features of this Schedule and in due course the Approved Programme, and should be established at an early stage of the project.

Except in the most simple of projects, the Plant is likely to consist of several Sections, some of which are operationally dependent on others. There will thus be a logical sequence in which completion of each will be required in order to achieve the most satisfactory commissioning of the whole plant. Furthermore the actual task of thoroughly inspecting and testing a completed Plant before it can be used safely can be costly in time and resources. To leave the whole of a project to be inspected and tested at the same time could lead to the procedure being rushed because of pressure from production management and hence lead to potential hazards, or else lead to unexpected delay if the inspection work is done properly. For this reason adequate time should be allowed for inspection and testing.

For each event for which the Purchaser requires to define the time-scale, this Schedule should specify:

(a) the time for commencing the event (either as a date, or by a period from the date for the commencement of the Works as defined in the Agreement or some other defined event); and

(b) the time for completing the event (by date or period).

Appropriate cross-references should be made to the relevant Schedules, in particular Schedules 14 (Criteria for the completion of construction), 15 (Take Over procedures) and 16 (Performance tests and procedures).

Where several Sections of the Plant or the Works are identified in the Specification or in Schedule 1 (Description of the Works) separate dates or periods should be stated for each.

In any event clarity is essential as to what is intended as an event for which a date or period is stated.

This Schedule should only contain events connected with liabilities for delay and other key events which are of importance to the Purchaser. Care should be taken not to restrict unnecessarily the Contractor's freedom to programme the work as appropriate.

In assessing the time or period for the Contractor to undertake the Works of a Section, or of the Plant as a whole, adequate time must be allowed to take account of the complexities of the Works, the time normally taken for customs clearance and other formalities connected with the importing of Materials and Contractor's Equipment, the degree of remoteness of the Site, climatic conditions and seasonal variations, and other relevant matters.

The identification in Schedule 11 of key dates or periods for completion of elements of the Works is in part linked, if the Contractor is late, to the application of liquidated damages for delay under Clause 15 (Damages for delay).

To avoid as far as possible the operation of Sub-clause 14.1, where the Contractor may become entitled to an extension of time, the events listed in this Schedule should be under the control of the Contractor without his having to rely on the performance of the Purchaser or the Project Manager. For this reason, events such as the 'Plant coming on stream' or the 'issue of an Acceptance Certificate' are not appropriate for the Contractor to have to achieve to avoid paying liquidated damages. Furthermore undue haste during the initial period of operation of a process may mean taking risks with the safety of the Plant and the personnel.

Sub-clauses 13.1 and 15.1 are worded in a way that places emphasis on the completion of construction as the event most appropriate for the application of liquidated damages for delay, but the Purchaser may prefer to apply such damages to a delay in Taking Over. If that is the case, the Purchaser should extend the list of criteria in Schedule 14 to include all the procedures he wants the Contractor to carry out ready for Taking Over by the Purchaser (see the check list in Guide Note J), and then write 'None' in Schedule 15 . Then the Project Manager could issue a Take Over Certificate under Sub-clause 33.7. Such an arrangement would avoid any confusion about how far take-over procedures had been completed by a certain date, which could arise as a result of the application of Sub-clauses 33.8 to 33.12 inclusive.

Whatever the event, it requires precise definition, preferably giving objective tests to be achieved rather than relying on subjective views of the Project Manager, to minimise the possibility of dispute as to the time when or if the event actually occurs.

The Performance Test Commencement Period, defined in Sub-clause 1.1 and referred to in Sub-clause 35.4 is to be stated in Schedule 11. This period is to run from Take Over, and is the period within which the Purchaser should commence any performance tests in accordance with Clause 35. Similarly the Performance Test Period, also defined in Sub-clause 1.1 and referred to in Sub-clause 35.6 is to be stated in Schedule 11. This period is to run from the commencement of the first performance test, and is the period within which the Purchaser should conduct the performance tests. Depending on the nature of the Plant this period may be just a few weeks or it may be as much as a year, particularly if the process depends upon, for example, availability of specified feedstock, is affected by ambient conditions which may vary over the year, or involves availability and reliability testing.

It should be noted that both the Performance Test Commencement Period and the Performance Test Period, even if extended in accordance with Sub-clause 35.4 or 35.6 as appropriate, must not extend past the end of the Defects Liability Period (see Sub-clause 35.7).

Schedule 11 is referred to in Sub-clauses 1.1, 4.1, 4.2, 13.1, 13.2, 15.1, 21.2, 21.9, 32.8, 41.9 and 42.3.

## Schedule 12: Liquidated damages for delay

*See Clauses 4 (Purchaser's responsibilities) and 15 (Damages for delay), Guide Note J (Completion, Taking Over, testing and start up) and Guide Note K (Incentives for early completion)*

This Schedule must clearly define the amount of liquidated damages for delay and the events to which they apply.

The events should correspond with those listed in Schedule 11 (Times of completion).

Liquidated damages are normally expressed as a sum of money or percentage of the Contract Price against a specified unit of time e.g. per week of delay. If the delay involves a partial unit of time, such as one week and three days, thought should be given as to how the part unit may be treated.

Liquidated damages should not exceed the genuine pre-estimate of loss that the Purchaser may suffer if the Contractor is late. A sum which is excessive is liable to be struck down in its entirety as a 'penalty' regardless of how it is described.

If the Purchaser is prepared to limit the Contractor's liability for delay (which may be reflected in lower Contractor's rates or fees), an upper limit for the total liquidated damages payable for delay may be stated. This may be expressed as a percentage of the Contract Price or as a monetary figure, which should be a multiple of the daily/weekly rate of damages so that the date when the maximum is reached can be easily ascertained. The Purchaser should note that, in accordance with Clause 15, the Contractor has no liability to pay further delay damages once an upper limit is reached. Nevertheless, depending on circumstances, in the event of delay beyond the maximum, the Purchaser may have grounds for termination under Clause 44 (Termination for Contractor's default), in which case he will have further remedies under that clause.

If there are to be no liquidated damages, the parties should be clear whether the intention is that there will be no damages for delay *at all*, or simply that there are no stated liquidated damages but that general damages for proven losses will apply instead. Under some contract strategies, in particular where there is an effort to align the interests of the Purchaser, the Contractor and possibly other contractors towards an overall project objective (sometimes called 'partnering'), no damages for delay are payable, whether liquidated or general. Instead the programme may be incentivised through bonuses.

If it is intended that liquidated damages for delay shall not apply but that general damages do apply, a Special Condition should state that the text of Clause 15 is deleted and replaced by a simple statement that general damages for delay shall apply. Schedule 12 should be marked as 'not used – general damages apply'.

If it is intended that there are to be no damages whatsoever for delay, a Special Condition should state that the text of Clause 15 is deleted and replaced by a simple statement that no damages for delay (whether liquidated, unliquidated or general) shall apply. Schedule 12 should be marked as 'not used – no damages payable for delay'.

See also Guide Note K.

Schedule 12 is referred to in Sub-clause 15.1.

## Schedule 13: Pre-installation tests and procedures

*See Clause 22 (Inspection and pre-installation tests)*

Clause 22 provides for the Project Manager or his nominee (see Sub-clause 22.8) to carry out off-site inspections and tests to satisfy himself that the work or Materials are in accordance with the Specification.

It is recommended that where possible the testing of Materials should be demonstrated in the manufacturer's premises by running factory acceptance tests. The items recommended for factory testing may include:

(a) key equipment items such as reactors or pressure vessels;
(b) machines;
(c) packaged equipment or prefabricated units;
(d) critical software and control systems.

The tests could include any appropriate form of test, such as pressure testing or load testing. The inspections could cover, for example, welding records or certificates of origin of alloys. Software tests could be devised by the Contractor/manufacturer to demonstrate that the software meets the Specification.

This Schedule should include:

(a) a description of the work, Materials or software to be tested;
(b) a description of the test and, where appropriate, details of the test rig, measurement accuracy, outputs to be measured and minimum duration of test; and
(c) the results to be achieved during each test together with acceptable tolerances.

Schedule 13 is referred to in Sub-clauses 22.2 and 22.4.

## Schedule 14: Criteria for the completion of construction

*See Clause 32 (Completion of construction) and Guide Note J (Completion, Taking Over, testing and start up)*

This Schedule should define the criteria to be satisfied by the inspection and testing of the various elements of the Plant, or any part of it, in accordance with Sub-clause 32.2.

Typical inspections and tests required to verify completion of construction are listed in Guide Note J.

Overall, the intention of the completion of construction inspections and tests is to demonstrate that every individual element of the Plant has been manufactured, constructed and erected/installed correctly prior to starting up the Plant.

If the Plant, or any part of the Plant, does not pass any inspection or test then the Project Manager should endorse the draft Construction Completion Certificate (see Sub-clause 32.7). The Project Manager should also endorse the draft Construction Completion Certificate if any matters of major importance are found during the inspections and tests, such as anything which would be likely to:

(a) cause a health or safety problem;
(b) cause an environmental issue;
(c) cause a loss of electrical supply or a trip;
(d) cause a breach in site or Plant security;
(e) contravene statutory legislation;
(f) restrict effective maintenance or operation of the Plant.

Note that Sub-clause 32.6 allows the Project Manager to issue a complete Construction Completion Certificate with a note of minor items to be completed before the issue of the Take Over Certificate. Such items usually include any work which is best left to near the end of the Works to ensure it is not damaged during other construction works, or other minor items that do not affect the operability of the Plant.

Schedule 14 is referred to in Sub-clauses 32.2 and 32.7.

## Schedule 15: Take Over procedures

*See Clause 33 (Taking Over) and Guide Note J (Completion, Taking Over, testing and start up)*

This Schedule should list any outstanding inspection and commissioning procedures as well as the take over tests themselves. Documentation by the Contractor and any approvals from third party organisations needed to demonstrate that the Plant may safely be started up, operated and maintained must be stated.

The parties may decide pre-contract that this Schedule cannot be satisfactorily completed until the Contractor has had time to carry out further design and engineering work, in which case a Special Condition should be included stating the time by which a draft of the complete Schedule shall be submitted by the Contractor together with the procedure for agreeing the final content of the Schedule.

It will be a matter of judgment for the Purchaser to decide how much of the work involved in finalising the preparations for start up should form part of the take-over procedures, when the Contractor is still in charge of the Plant, and how much should be done after the Purchaser has Taken Over the Plant. For example, the task of charging catalysts could be done by the Contractor, with the assistance of the Purchaser's staff, but charging raw materials, process chemicals and fuel could be done by the Purchaser after Taking Over.

It is normally the case that all operating chemicals, raw materials, feedstocks and utilities will be provided by the Purchaser. This should be stated in Schedule 15.

Schedule 15 is referred to in Sub-clauses 33.2 and 33.7.

## Schedule 16: Performance tests and procedures

*See Clause 35 (Performance tests), Schedule 17 (Performance guarantees and damages for failure), Guide Note J (Completion, Taking Over, testing and start-up) and Appendix A (Performance Test Flowcharts)*

### Clause 35

Clause 35 applies to the Contract if the Contractor has made specific guarantees in respect of the performance of the Plant verifiable by performance tests (see Sub-clause 35.1).

Guarantees of performance measured by tests are the mechanism used to establish how well the Plant is performing against the Contract requirements. If the guaranteed level of performance is not achieved, it is usual to provide for compensation in the form of damages payable to the Purchaser for the lower-than-expected (and promised) performance. In some cases, a bonus may be payable to the Contractor for better than guaranteed performance. If this is to apply, a Special Condition must deal with the point.

Schedule 16 (see Sub-clause 35.3) should describe the performance tests that are to be carried out, their duration, the raw materials to be used, what product is to be made, the conditions under which they are to be carried out, how the Plant is to be operated and so on. The parameters that are actually to be guaranteed and their associated liquidated damages are then set out in Schedule 17.

The first part of Clause 35 provides a clear structure for the carrying out of the performance tests.

Sub-clause 35.4 provides for a defined period within which the Purchaser is to commence the performance tests (the Performance Test Commencement Period), the initial length of which must be stated in Schedule 11 (Times of completion). The Sub-clause also deals with the situation where not all performance tests have been commenced within the Performance Test Commencement Period.

Sub-clause 35.6 requires the performance tests to be carried out within the Performance Test Period, the length of which must be stated in Schedule 11, and deals with the possible extension of that Period. The Performance Test Period commences with the commencement of the first performance test. In no circumstance whatsoever may the Performance Test Period extend beyond the end of the Defects Liability Period (Sub-clause 35.7).

If a performance test is failed, Sub-clause 35.11 provides for the Contractor to be allowed access to the Plant to make adjustments or modifications to improve its performance. He may at his discretion go through several cycles of adjustments and tests provided he does not run out of time at the end of the Performance Test Period – including time for the last performance test to demonstrate any improvement made in the Plant's performance.

The structure for the performance tests created by Clause 35 is shown in the flowcharts in Appendix A. These clearly show that there must be careful consideration of the balance between the Performance Test Commencement Period and the Performance Test Period, as stated in Schedule 11, and the Defects Liability Period. Not having enough time to set up and run the test envisaged or the combined time of the test commencement period and the test period being equal or nearly equal to the Defects Liability Period could have a detrimental effect on what the parties intend – see Sub-clause 35.7.

In certain situations it may be appropriate to require that performance tests which do not directly relate to performance guarantees are undertaken. Clause 35 does not apply to such tests by virtue of the terms of Sub-clause 35.1. However, whether included within the original scope of the Works as stated in the Contract, or ordered as a Variation, there is no reason why the Contractor should not assist the Purchaser with the conduct of such tests. Their contractual significance would depend on what errors or lack of performance they demonstrated, including for example a failure to comply with Sub-clause 3.4 or a Defect requiring correction under Clause 37 (Liability for Defects).

### Performance tests

If performance tests and guarantees are to be included, the drafting of Schedule 16 will need considerable care. It is probably the most difficult Schedule to prepare well, and commonly the most neglected. At the time of drawing up the Contract performance tests seem far away and difficult to define in detail. But careful attention to these details pays dividends by avoiding discord on site at the critical time of performance testing.

Schedule 16 should specify how the Plant is to be prepared for the tests, the duration of tests, the performance criteria to be measured, conditions governing interruptions, methods of measurement and analysis, (which should be standard and widely used whenever possible), and measuring tolerances to be allowed to cover lack of precision. Routines for measuring stocks at start and finish, recording and interpreting instrument readings and collecting preparing and analysing samples should all be included as necessary.

Many of these requirements may affect the detailed design of the Plant. For example, a performance test may require measurement of the flow of a stream of solids to an accuracy of one percent when for normal operation an accuracy of four or five per cent may be sufficient. This might make batch weighing necessary instead of, or as well as, continuous belt weighing during the test.

Schedule 16 should also state how the measuring tolerances should be allowed for in calculating the results of the tests for comparison with the guaranteed values.

The Plant will usually be required to operate satisfactorily over a range of conditions. It is probable that corrections to the guarantees will need to be applied if the performance test takes place, as frequently occurs, under conditions differing from the basic design case. For example, the guaranteed capacity of a Plant may equal the nominal capacity when the ambient air is at the basic design temperature, but an adjustment factor may be necessary to take account of differences between actual temperature during the test and the design temperature.

Performance criteria to be tested may include Plant capacity, details of consumption and quality of raw materials and utilities and essential criteria for the quality of the finished products. In some cases guarantees may also be required for reliability, the consumption of general utilities if these are a significant cost factor and the quantity and quality of by-products (including steam and condensate) and polluting effluents.

The problem for anyone designing a performance test is that there are so many variables. A test might have a duration of an hour, a shift, a day, a week or a month. The test might cover one product, or two or several. It might seek to measure one critical parameter, two, three or more. Essentially the task is to be completely practical and to recognise that not every element of performance can be tested.

**Unspecified performance tests**

Sub-clause 35.3 also provides for possible tests to be carried out that have not been specified in Schedule 16. If an agreement is reached between the Project Manager and the Contractor that any performance tests are to be carried out which are not specified in the Schedule they should be treated as Variations. Care must be taken when defining the scope and purpose of any new or unforeseen tests since they may involve an addition to the performance criteria already set out in the Schedule.

It should also be noted that such tests might necessitate alterations to the original Plant design to enable additional temporary or permanent instrumentation and sample points to be installed. The requirement for both parties to agree that the tests should be carried out is an appropriate method of proceeding, particularly if the timing of a request to carry out such tests would not allow the work to be carried out for reasonable time or cost. The Contractor should be in a position to decline to carry out such tests particularly if he is likely to incur overall delay, irrecoverable costs or additional liability. There should be no implication that such tests should be part of the Contractor's obligations unless required for statutory reasons.

Ideally, additional performance tests should be avoided and every effort should be made to define the full extent of the tests required at the pre-contract stage.

Schedule 16 is referred to in Sub-clauses 35.3, 35.8 and 35.10.

# Schedule 17: Performance guarantees and damages for failure

*See Clause 35 (Performance tests), Schedules 13 (Pre-installation tests and procedures) and 16 (Performance tests and procedures), Guide Note J (Completion, Taking Over, testing and start up) and Appendix A (Performance Test Flowcharts)*

**Schedule 17**

For any performance guarantee to apply, it must be stated explicitly in Schedule 17 at the time of signing the Contract or by means of a subsequent agreement as described in the guidance on compiling Schedule 16 (under the heading 'Unspecified performance tests'). It cannot simply be implied or inferred. For each guaranteed parameter (quantity or quality of product or effluent, usage of raw material, utility etc.), Schedule 17 should include:

(a) guaranteed performance;
(b) maximum permissible deviation; and
(c) the amount of money to be deducted for each unit of shortfall as liquidated damages.

The damages arising from a failure to achieve the guaranteed performance should wherever possible be established in advance as liquidated, "agreed", damages. IChemE's strong recommendation, reflecting good industry practice, is to allow for a clearly defined margin of failure of the guarantee or guarantees to be covered by liquidated damages. Such liquidated damages are intended as agreed compensation for the Purchaser's increased operating costs, loss of product, etc, as a result of the Plant's poorer performance, and in theory should be no more than a genuine pre-estimate of the loss which would be suffered by the Purchaser in operating the Plant under such conditions in practice.

The parties should note that if the agreed contractual position is that the Contractor is to be liable for any other damages, then this should be explicitly stated in this Schedule or in a Special Condition.

Though it is not practicable here to make detailed recommendations the following may be a useful aide-memoire.

(a) *Raw material consumptions, utility consumptions, yields, efficiencies*
These represent direct running costs. Liquidated damages could be set at a figure equalling the extra costs for a certain period of time. Figures between one and three years at flowsheet rate have been used.

(b) *Groupings of different running costs*
In general, savings in one item should be offset against excess costs of another. Occasionally, however, this is not appropriate—for example, when a particular utility or feed material is in limited supply.

(c) *Credit for by-products*
Useful by-products would normally be grouped together with the input materials, the total running costs being arrived at mathematically, but there are instances when a by-product may become a nuisance if made at an increased rate.

(d) *Plant capacity*
A true appraisal of the Purchaser's financial loss through shortage of capacity is very difficult to make, as it is affected by his ability to make full use of the capacity available. Note that losses of this nature will largely fall within the exclusion of liability stated in Sub-clause 45.1 unless included within liquidated damages.

(e) *Product quality*
It is sometimes possible to relate the value of a substance to its purity and to set liquidated damages for excess operating costs. If there were two or more products, their inter-relation should be considered and credits in the quality of one product set off against shortcomings in another. Maximum permitted limits of impurities may, however, be used to impose restrictions.

(f) *Effluent quantity and quality*
Excesses in either may result in higher disposal costs, to which liquidated damages could be related. Absolute limits may be imposed by statutory authorities or other bodies.

(g) *Groupings of performance criteria*
It is important to consider whether, and the extent to which, different criteria of performance should be taken together. For instance, a Plant may fall slightly short of the promised output, but be more efficient; another Plant may make a product to the desired specification and at more than the guaranteed capacity but with excessive energy consumption.

Performance guarantees should only apply to really essential parameters. Liquidated damages should be unambiguous with respect to any grouping of criteria and should be defined in a way that makes their calculation simple and clearly defined.

An upper limit of total liquidated damages relating to plant performance will normally be set. Any such upper limit to the total liquidated damages must be set in the context of the Contractor's overall aggregate liability under Sub-clause 45.3, and it must also be clear whether any such upper limit includes or excludes liquidated damages for delay as well as liquidated damages for failure of performance. If there is any inconsistency, for example the upper limit on liquidated damages is greater than the total aggregate liability under Sub-clause 45.3, then the total aggregate liability in respect of Sub-clause 45.3 and stated in the Form of Agreement will take precedence over any upper limit of liquidated damages stated in Schedule 17.

**Clause 35**

While the first part of Clause 35 sets out the framework for the conduct of the performance tests, as described in the guidance on Schedule 16, Sub-clauses 35.12 – 35.15 deal with the situation where the Plant has failed a performance test and damages become due. The Clause also addresses the situation where the performance of the Plant falls substantially outside the range over which liquidated damages apply.

It will not always be possible to establish a defined margin of failure (e.g. exceeding the guaranteed values by 10%) for all performance tests. In some circumstances the required guaranteed performance or quality will be absolute (e.g. product quality as required by regulatory or safety requirements). Therefore any failure will be unacceptable. In such circumstances the only remedy might, for example, be to operate the Plant at a reduced output and thereby achieve the required quality. Again if at all possible the damages should be stated as liquidated damages by reference to the reduced output of the Plant at which it must be operated in order to deliver the required quality of product.

IChemE's view is that in all normal circumstances liquidated damages should provide the complete remedy for failure to meet guarantees of performance. This is important as it enables the Contractor to assess his risk when tendering and achieves finality between the parties. In contrast general damages do not allow the Contractor to assess his risk with any certainty. However, if the Plant performance is very substantially outside the guaranteed level or where it is not possible either to modify the Plant or to change its operating conditions to achieve an absolute product or discharge quality standard then in such circumstances, if not already covered by a Special Condition, general damages or even rejection of the Plant as a whole may be the only remedy.

If it is the intention of the Parties that general damages should apply as well as liquidated damages, the level of performance at which the change from liquidated to general damages should occur should be stated in Schedule17. It is this changeover point which is referred to in Sub-clauses 35.12 – 35.14 as "any limit for liquidated damages". If it is the case that the required performance is absolute and liquidated damages are inappropriate, a clear statement should be made in Schedule 17 that liquidated damages will not apply to any failure to achieve the absolute guarantee. It will also be necessary to give consideration as to whether, in these circumstances, the plant is in reality usable, and if not, what the consequences should be.

Many users may wish to include an added refinement of having a limit to the extent of liquidated damages but with general damages not applying unless the poor performance is significantly worse than that represented by the maximum amount of liquidated damages. Such a situation may be dealt with either in Schedule 17 or by a Special Condition.

Note that if general damages are to apply they will be subject to Sub-clause 45.1 which excludes various stated categories of damages many of which would fall into the category of indirect or consequential damages. Sub-clause 45.1 does not apply to liquidated damages (see sub-paragraph (iii) of Sub-clause 45.1) as liquidated damages provide the opportunity for the Contractor to know his risk exposure in a way that general damages do not. In any event it may be that the amount of liquidated damages has to be reduced to a value which is commercially acceptable.

It is also worth noting that Sub-clause 45.3 provides for the Contractor's total aggregate liability under the Contract to be capped: the Purchaser should consider carefully how any stated limit for the application of liquidated damages interacts with such cap on the Contractor's total aggregate liability.

In some circumstances the failure of a performance guarantee may be linked directly to the failure of a subcontract element of the Plant. If the performance of a subcontract element of the Plant is critical to the performance of the whole Plant it would be appropriate to consider whether a Special Condition should be included requiring the Contractor to include in the subcontract for the Purchaser to be able to exercise third party rights against the Subcontractor or to require the Contractor to obtain a collateral warranty in favour of the Purchaser from the Subcontractor. If the former approach is adopted and if the subcontract is to be based on the IChemE Yellow Book, then a suitable amendment would be necessary to Sub-clause 9.8 in the Yellow Book.

Sub-clause 35.12 deals with the situation where the performance test has failed within the Performance Test Period. If the failure is within any limit specified in Schedule 17 then liquidated damages for failure will apply.

If the test failure lies outside such a limit for liquidated damages, under Sub-clause 35.13 the Purchaser may, but is not bound so to do, instruct the Project Manager to extend the Performance Test Period and to instruct the Contractor as to what actions he should take.

Whether or not the Purchaser acts in accordance with Sub-clause 35.13, Sub-clause 35.14 provides for the Project

Manager to determine the amount of compensation due to the Purchaser if the performance test results are outside any limit for liquidated damages.

Sub-clause 35.15 provides for the situation where for whatever reason a performance test has not been completed, or the Contractor has made modifications or adjustments to the Plant after a failed performance test, but then time runs out and a repeat test is either not done at all or is only partially carried out. The question then is what to do, having regard to the fact that Sub-clauses 35.12 and 35.14 apply to the results of completed tests. In such a situation Sub-clause 35.15 requires the Project Manager to determine the compensation due from the Contractor to the Purchaser, making use of the results from any completed or partially completed tests and all other relevant information and circumstances. This gives the Project Manager wide powers. In making his assessment he must be mindful of his obligations under Sub-clause 11.1(d). If there are almost no meaningful performance test results for the Project Manager to consider, but the Plant is generally performing well, he might think it appropriate to determine that the Plant has passed its tests.

Finally, if, for reasons outside the Contractor's control, a performance test has not been commenced at all in time for it to be completed by the end of the Defect Liability Period the test will be deemed to have been passed (Sub-clause 35.16).

Schedule 17 is referred to in Sub-clauses 35.1, 35.12, 35.13, 35.14 and 35.17.

## Schedule 18: Valuation of Variations and claims

*See Clause 19 (Claims) and Guide Note F (Variations and claims)*

It should be noted that there are several pricing mechanisms that can be used in the valuation of variations including lump sums, fixed costs, measured rates or combinations of all of these. It is important that the Contract is clear which method or combination of methods is to be used in order to avoid misunderstanding and potential conflict later.

In all other situations of valuing Variations and in valuing claims where respectively Sub-clause 18.3 and Clause 19 apply, the following approach is recommended, in which the valuation is based on costs incurred and Schedule 18 provides tables for the Contractor to state, at the time of tendering, his unit rates for the various items described. An opening statement in Schedule 18 might be made along the following lines:

> 'In the absence of the acceptance of a quotation in accordance with Sub-clause 18.2, the method of establishing the Cost of carrying out any Variation or the Cost which is the subject of a claim should be on the basis of any rates stated in this Schedule, and otherwise the itemised net costs of Materials, Contractor's Equipment and personnel, to which charges may be added according to any applicable tables shown below, together with the net cost of any further items not covered by any rates stated in this Schedule.'

The tables to be included in Schedule 18 may include the following, and provide the tenderer with the opportunity to include his unit cost rates:

(a) *Rates for home office personnel*
A typical table would list the rates per hour for the various grades of personnel that are chargeable under the Contract. It is important to specify the rates that are payable for overtime working (if applicable), and to state what costs are included in the overheads (see (b), (c) and (h) below).

(b) *Reprographic rates*
A typical table of reprographic charges would specify the rates applicable to different types of reprographic processes. If the rates vary according to the number and size of the copies, this should be stated.

If reprographic costs are not separately chargeable, but are covered elsewhere, for example, in the hourly rates, or in a fixed fee, it should be made clear that the Contractor is only obliged to supply a certain number of copies of any documents to the Purchaser and this number should be specified in Schedule 18.

(c) *IT systems*
If not included within the personnel hourly rates, separate rates for PCs, CAD and other IT systems and any bought-in computer time should be stated both for hardware and software.

(d) *Rates for site office and supervisory personnel and site labour*
As for (a) above, except that rates are normally quoted on a daily or weekly basis. If weekly rates are quoted, the table must specify the number of working days per week. Clarity is also needed concerning overtime payments (if any).

If local living allowances are payable in addition to daily or weekly rates, details must be given.

A percentage is often charged on the cost of site labour to cover such elements as:

(i) small tools and consumables (see (f) below);
(ii) statutory charges (for example, training levies); and
(iii) all holiday, sick and other such 'non-productive time' pay.

(e) *Travelling and subsistence allowances*
Travel time is normally chargeable at the appropriate hourly, daily or weekly rates, but this is not always the case and the position must be clearly defined.

If any allowances are paid for travelling and associated subsistence instead of reimbursement at net cost incurred, this must be made clear and the rates given.

(f) *Contractor's Equipment*
If it is intended to use the Contractor's own tools and equipment, a table of unit rates should be agreed. Small tools and consumables are normally included in the site labour rate (see (d) above). Hired Contractor's Equipment will be paid for on the basis of the hire charges incurred.

(g) *Procurement rates*
If procurement services are not chargeable on the basis of hourly rates, they may be chargeable as percentages of the value of materials and services bought and subcontracts placed. Such percentages must be quoted.

It should be noted that a rate for an item (especially a generic item such as personnel of a given grade) may not correspond precisely to the net cost actually incurred by the Contractor for the item. The Contractor will take the benefit or loss on the agreed rate. The term 'cost rate' (or similar) should not be used to describe agreed rates because it may give rise to a dispute whether the agreed rate can be re-opened and adjusted against net cost actually incurred.

In preparing the tables for Schedule 18 as described above it should be clearly understood that any rate is not to include profit. Profit is a defined term in the Red Book with its value as a percentage addition to Cost stated in the Agreement. Sub-clauses 18.3, 19.4 and 43.5(c) refer to the addition of Profit.

In addition to charges calculated from the unit rates given in paragraphs (a) to (g) above, there may be one or more fixed fee elements covering, for example, overheads and the supply of certain services. If the fixed fee is to include an element for profit it should be clearly stated how that interacts with the percentage Profit addition to the Cost of the work as described in (a) to (g) above. Any such fees must be quoted, and it must be clear which cost elements are covered by such fees. In other cases, there may be no fixed fee, but there may be a percentage fee based on the total reimbursable cost.

If any of the rates quoted in the tables are valid only up to a certain date, this must be made clear and the basis of revision stated.

Schedule 18 is referred to in Sub-clauses 18.1 and 19.3.

## Schedule 19: Contract Price and payment

*See Clause 41 (Payment) and Guide Note Q (Bonds and guarantees)*

### Contract Price

The Contract Price may be stated in one or more currencies. The total values which form the Contract Price in the various currencies are to be stated in Paragraph 3 of the Agreement. Schedule 19, under the sub-heading of Contract Price, will need to give the detail. At the very least it will be necessary to be clear as to which work and Materials are included within each part of the Contract Price, for example clarity between the provision to a port of imported Materials and the local transport costs in taking those Materials to the Site. (See also Guide Note G (Delivery to Site)).

### Price escalation

It may be appropriate to include a price escalation mechanism in respect of elements of the Contract Price, such as where the price of an element is largely dependent on the price of a specific commodity. If possible such a mechanism should be linked to published inflation indices, such that if the indices vary then the Price, or the particular interim payment to be made, varies according to the change in the index without further negotiation. If an appropriate index exists then clauses which make the application of the formulae dependent upon the Contractor actually experiencing additional costs, to be evaluated by using the indices, simply make for unnecessary work and potential grounds for disputes.

### Provisional sums

The Purchaser may wish to include in the Contract specific provisional sums which will be either:

(a) Monies in respect of work which will be required, but its cost has yet to be established – this is particularly relevant where it is intended that the work be undertaken by a nominated Subcontractor; or

(b) Monies in respect of work which may or may not be required, and for which an allowance is included in the Contract Price to facilitate the subsequent ordering of such work.

In either circumstance it would be appropriate to include a Special Condition in the Conditions of Contract, as well as identify the provisional item or additional work in Schedule 18.

An appropriate Special Condition would be as follows.

---

**EXAMPLE CLAUSE**

**x.1** If a provisional sum is specified in Schedule 19 in respect of any contingency or in respect of work which may be required, the sum shall only be expended if the **Project Manager** issues an appropriate instruction by means of a **Variation Order**.

**x.2** If a prime cost sum is specified in Schedule 19 for work to be done or **Materials** to be supplied by a nominated **Subcontractor** then, subject to the **Project Manager's** right to omit such work or **Materials** in pursuance of Clause 16 (Variations), the **Project Manager** shall within a reasonable time give the **Contractor** instructions nominating the **Subcontractor** and, subject to Clause 10 (Nominated Subcontractor), the **Contractor** shall place a subcontract in accordance with such instructions.

**x.3** In the final settlement of accounts all provisional and prime cost sums stated in Schedule 19 shall be deducted from the **Contract Price.** The **Cost** incurred by the **Contractor** in complying with the instructions issued under Sub-clause x.1 and in respect of nominated **Subcontractors** shall constitute additions to the **Contract Price**.

---

### Payment

Clause 41 provides for the payment of the Contract Price in instalments according to Schedule 19 and deals with the procedure to be adopted for the payment of each instalment. The General Conditions do not specify the pattern of the instalments or how the amounts to be paid are determined. These should therefore be set out in this Schedule.

Sub-clause 41.2 provides that payments to be made according to the completion of defined tasks, sometimes known as 'stage' or 'milestone' payments, are to be made only upon completion of those tasks. These must be clearly defined so as to minimise arguments about whether or not completion has been achieved. It is important to avoid events such as the issue of the Acceptance Certificates as the point for a stage payment, because they depend upon events outside the Contractor's control.

Examples of suitable events are:

(a) issue of Health and Safety Plan by the Contractor;

(b) issue of first complete set of process flow diagrams and relevant data sheets by the Contractor;

(c) completion of clearance and levelling of the Site in conformity with the approved design and specification;

(d) completion of foundations for a major element of the Plant in conformity with the approved design and specification;

(e) delivery to Site of a major item of equipment, complete as ordered, but not necessarily including spare parts;

(f) completion of erection of steelwork in conformity with the approved design and specification;

(g) completion of erection of defined Sections of the Plant in conformity with the approved design and specification;
(h) issue of the applicable Construction Completion Certificate for a defined major part of the Plant;
(i) issue of the last Take-Over Certificate;
(j) issue of the last Final Certificate.

Note the identification of discrete elements of the Works. The Contractor will most likely be employing different Subcontractors for each element and separate payments will facilitate his cash flow. A general guide is that the list of milestone achievements should be sufficiently detailed to allow the Contractor flexibility of programming and good cash flow. Note that a sound approach is that project cash flow should be as nearly as possible neutral between the parties, that is that neither is financing the other during the progress of the Works. As the exact nature of the Works may not be known until after the Contractor has submitted his tender, it is almost inevitable that the detail of this Schedule will need to be finalised after tenders have been analysed and before the Contract is awarded. There is a good argument for setting out guidelines in the Instructions to Tenderers and inviting tenderers to propose many of the milestones, with the inclusion of those events stated above.

It is worth noting that milestone payments against issue of Take-Over and Final Certificates have a similar effect as withholding a retention which is then released at these two stages.

Clause 41 complies with the amendments to the Housing Grants, Construction and Regeneration Act 1996 set out in Part 8 of the Local Democracy, Economic Development and Construction Act 2009, which came into force in England and Wales on 1 October 2011 and in Scotland on 1 November 2011. One consequence of making the Clause compliant with the statutory changes is to remove the earlier approach of the Contractor submitting an invoice to the Purchaser after the issue of the Project Manager's certificate. In that respect the Red Book is now the same as the Green and Burgundy Books.

It should be noted that, in accordance with legislation, if no notification (i.e. no certificate) is issued by the Project Manager in accordance with Sub-clause 41.4 then, subject only to Sub-clause 41.7, the whole of the Contractor's request for payment submitted in accordance with Sub-clause 41.3 becomes payable – see Sub-clause 41.6. **Purchasers and Project Managers should note the importance of this, and note that if for whatever reason no monies are in fact due to the Contractor then the Project Manager's certificate must still be issued and must state that the sum due is zero.**

If no certificate is issued, then the Purchaser has one further opportunity to correct the position, by issuing a 'pay less' notice in accordance with Sub-clause 41.7. Failure to issue the pay less notice where no certificate is issued means that the whole of the Contractor's request for payment is payable. More generally the pay-less notice is a means whereby the Purchaser can notify an intention to pay less than the full amount certified, for example where liquidated damages or other counter-charges are due from the Contractor.

As noted above, clarity will be needed if payment is to be made in more than one currency as to which currencies will apply to payments for the various elements of the Works and the Plant, and how any price escalation provision is to work for both interim and final payments.

Schedule 19 is referred to in Sub-clauses 41.1 and 41.2.

## Schedule 20: Contract co-ordination

*See Sub-clauses 3.7, 5.4 and 11.8, and Schedule 1 (Description of the Works*

A contract co-ordination procedure is essential if the parties are to carry out their duties effectively, efficiently and in accordance with the project programme. The drafting of Schedule 20 is therefore of great importance, although on some occasions the parties may decide to draft and approve a separate document sometimes referred to as a Project Procedure. The Schedule should include the following:

(a) primary points of contact in each party and the Project Manager including addresses, telephone, facsimile and email addresses;
(b) specific methods of identifying documents including sequential numbering systems;
(c) a common project index for numbering and filing documents in electronic and hard copy format;
(d) methods of document and data transfer, including email or secure email;
(e) arrangements for the issuing of invoices to meet the respective parties' commercial operating standards and methods;
(f) methods of transferring payments between the parties including electronic banking details.

It is emphasised that informal communications, whilst helpful to the day to day running of the project, must at all times be confirmed in writing so that accurate instructions or requests for information can be distributed to all relevant members of the Purchaser's and Contractor's teams.

Attention is drawn to Sub-clauses 1.6, 5.1 and 11.8. If it is intended that the contents of Schedule 20 are inconsistent with either Sub-clause a Special Condition to amend the General Conditions will be necessary.

Schedule 20 is referred to in Sub-clause 5.4.

## Schedule 21: Reports and records

*See Sub-clauses 3.7, 16.8 and 39.3*

### Reports

Sub-clause 3.7 requires the Contractor to provide a monthly report to the Project Manager on the progress of the Works, supporting his reports with appropriate documents including any update of the Approved Programme. Schedule 21 is the appropriate place to detail any more specific requirements of that general obligation, itemising any particular requirements or topics which the monthly report should contain or the format in which it should be provided. Where the Project Manager has in turn to provide a report on the progress of the Works in a particular format he may wish the Contractor to report in the same style.

In detailing the requirements of the reports to be provided in accordance with Sub-clause 3.7 consideration should be given to the Contractor being required to report progress against the Approved Programme using a recognised critical path analysis package, together with an executive summary.

 

**Records**

Sub-clause 3.7 requires the Contractor to provide and maintain records as detailed in Schedule 21. Such records should include details of progress and of the labour, Contractor's Equipment, Materials and Subcontractors on Site. These requirements should be briefly set out in this Schedule.

Clause 40 (Records and audits) sets out the contract requirements for the length of time that records are to be kept.

It should be noted that records may well be of technical matters and not simply costs, and again Schedule 21 should set out the Purchaser's requirements of such records.

When the Contractor considers that he is entitled to claim extra time and/or money he will need to keep more detailed records of the circumstances, actions taken, equipment and Materials used, and similar details. The Purchaser and the Project Manager will need to consider the extent to which it is necessary to define the requirements for such additional records. It should be noted that records are increasingly being held in electronic rather than paper format.

The Schedule should contain, as far as possible, samples of the forms to be used, or in the case of computerised records, printouts of typical records.

In any case, whatever form the records are to take, there will be certain basic requirements. Each document should show:

(a) contract title and identification number;
(b) topic covered, identified in relation to the contract documentation;
(c) date on which the work was carried out or on which the expenditure was incurred;
(d) name of person working, hours worked and resources used by him, as appropriate;
(e) quantities of materials or other resources used, if applicable; and
(f) in the case of subcontracted work or purchases from suppliers, the invoiced cost.

Schedule 21 is referred to in Sub-clauses 3.7 and 40.1.

# Guidance on contractual issues

## Guide Note A: Innovative work or checks by Purchaser, process licensor or Contractor

If there is a large innovative element in the design, especially in the process design, for the Plant then it may be preferable to use a reimbursable form of contract. This is because when the Contractor is limited by a lump sum contract he will be reluctant, or may simply not be able, to explore all the possibilities of optimisation or innovation in design. The reader is referred to the IChemE Form of Contract for reimbursable contracts (the 'Green Book') or the target cost form (the 'Burgundy Book').

Even where there is a significant innovative element, it will often be the case that a considerable part of the Plant will be well defined, so that the Contractor can bid for that part on a lump sum basis. In that case it is the responsibility of both parties to define clearly the elements of the work which are to be covered by the lump sum, as opposed to those elements which may need to be covered by a provisional sum or to be paid for on a reimbursable basis because they cannot be properly specified at the time of contract.

The Purchaser may require the Contractor to carry out innovative design work perhaps for a particular section or part of the Plant where he wishes to make his own technical input but does not want to impose any design in the contract. Examples might be the possible use of catalysts outside their normal operating conditions, or the use of new and previously untried materials in fabrication. In these cases it would be sensible to include a Special Condition in the Contract, perhaps giving the Purchaser the right to review the Contractor's design before the formal submission of Documentation for Approval with Schedule 2 (Documentation).

For example:

---

**EXAMPLE CLAUSE**

## XX. Purchaser's right to review design work

**XX.1** The **Purchaser** will require to review the design work carried out by the **Contractor** relating to those parts of the **Plant** or aspects of the design of the **Plant** so designated in the **Specification**. The **Contractor** shall keep the **Purchaser** informed of the progress of such design work by the preparation of reports at appropriate stages and in any event at intervals of not more than thirty days starting from the commencement of such work. The **Contractor** shall afford reasonable access during normal working hours for the **Project Manager** or his designated representatives to the personnel carrying out such work, and shall provide the **Project Manager** with a copy of any document relating to such work, at the request of the **Project Manager** or his designated representative.

---

The Purchaser may also choose to retain an option, exercisable when the Contractor has completed the design work on Sections covered by provisional sums, of having an agreed price for such Sections of the Plant incorporated in the lump sum through to completion of the project, in which case an appropriate Special Condition should be included at the outset. It may also be necessary to have a Special Condition to clearly establish the guarantees and liabi ities applying to this innovative design.

In the case of technology licensed from a third party, whether by the Purchaser directly or by the Contractor as part of his Contract, particular care must be taken to define in a Special Condition and in such Schedules as 1, 2, 3, 6, 9, 13,15 and 16 any role the licensor's personnel may have in checking the Contractor's design documentation, the provision of guarantees and advice and interpretation of process and engineering information provided as part of the licensor's documentation or their role in other phases of the project such as training, commissioning or performance testing.

The Purchaser may require the Contractor to check information provided by the Purchaser before he uses it, or may wish to rely on some special experience of the Contractor in designing and constructing a particular type of plant.

Where the Contractor's special experience of the technology is involved, a suitable wording for a Special Condition might be as follows:

---

**EXAMPLE CLAUSE**

## XX. Contractor's warranty

**XX.1** The **Contractor** warrants that he is qualified and experienced in the design and construction of plants involving technology of the type to be used in the **Plant**. The **Contractor** shall apply his expertise in this respect to the examination of the process design information supplied to him by the **Purchaser**, and shall promptly bring to the attention of the **Project Manager** any error, omission or inconsistency in such information of which he becomes aware. The **Contractor** shall assist the **Purchaser** to remedy such error, omission or inconsistency and to improve the process design information to a suitable extent, provided that in so doing he shall not be required to breach any undertaking given to a third party in respect of proprietary information.

---

Any changes to the Specification of the Plant resulting from the Contractor's review of the process design should be treated as a Variation in accordance with Clauses 16 (Variations), 17 (Variations proposed by Contractor) and 18 (Valuation of Variations) of the General Conditions.

# Guide Note B: Intellectual property, know-how, confidentiality, information and software

*See Sub-clauses 4.1 and 43.3 and Clauses 8 (Patent and other protected rights) and 20 (Confidentiality)*

### Intellectual property

Intellectual property is a valuable asset. The Purchaser and Contractor should identify and protect it by appropriate measures. Various aspects of protecting intellectual property are reviewed below.

### Process technology sources

If the Plant is to be based on third party technology as previously arranged by the Purchaser, the Purchaser will have entered into a licensing agreement with that party, including confidentiality provisions and patent protection arrangements. The confidentiality provisions of the licensor will also apply to the contractors during the tendering stage and require separate Confidentiality Agreements to be concluded prior to the issue of tender enquiry documents.

If the design of the Plant is to be based on the Purchaser's own technology he will require confidentiality protection; Clause 20 deals with this. He will have already made his own IPR protection arrangements.

If the design of the Plant is to be based on the Contractor's technology there will have been a Confidentiality Agreement executed between the parties early in the negotiation process well before the preparation of tenders. This may be sufficient to last the duration of the Contract or may be replaced with a Confidentiality and Licensing Agreement which will also cover IPR matters. In either case they will remain in force after the completion of the Contract. Such agreements will need to be cross-referenced in this Contract. Clause 20 may need amending if a Confidentiality and Licensing Agreement is used. Duplication and conflict of clauses are to be avoided.

In many processes it is necessary to include some sub-processes which are licensed or purchased from third parties. This can range from a complete process package, requiring separate technology agreements, sometimes involving the supply of specialist equipment, to the simple purchase of proprietary equipment where the supplier states the necessary operating conditions and provides guarantees.

In all cases care must be taken to cover confidentiality, patents, and ownership of documentation.

When preparing final documentation (see Schedule 2 (Documentation)) it should be noted that confidential information would not normally be provided to the Purchaser unless agreed in a Special Condition or elsewhere.

### Confidentiality

Ideas protected by methods such as patents, copyright, registered design, design rights or trademarks may only constitute a small amount of an organisation's valuable information. The rest of an organisation's knowledge will simply be know-how. All the parties involved in a process project will have knowledge that they wish to be kept confidential.

The Contractor and Subcontractors will have confidential manufacturing knowledge and process know-how. The Purchaser will have confidential information about his own operations and his own processes. All parties will have confidential information of a commercial nature. They all have a considerable interest in protecting the confidentiality of their information.

Sometimes none of the above methods of protection are suitable. This may be because the know-how does not fit under any of them, or because one of the parties considers that secrecy is a better commercial protection. It is then necessary to require all individuals who need to know about the product or process to be bound by confidentiality agreements. This could include not only employees but also contractors and consultants and their employees.

The confidentiality obligations of both parties are set out in Clause 20 which together with Clause 8 defines the limitations governing the use by the parties of restricted information and Documentation obtained from the other. This covers both parties' own specialist designs and also any know-how which may derive from a process licensor or other outside source.

Sub-clause 8.8 allows the Purchaser complete freedom to use information in whatever manner serves the operation of the Plant, including the removal of bottlenecks, maximising output through minor changes and improving efficiency or yields. On the other hand it does not permit the Purchaser to build another plant like, or similar to, the Plant supplied under the Contract and based on the Documentation supplied by the Contractor. In between these extremes, however, it is possible to visualise various major additions that could result in increases in capacity or other improvements. Under Sub-clause 8.8 the Purchaser's right to carry out such extensions with the aid of the Contractor's designs is restricted to a capacity increase of 25 per cent. While these arrangements are suitable in a good many instances, there are cases in which the Purchaser's rights in this respect should not be limited at all, for example, if the process design is the Purchaser's own and the Contractor only provides an engineering service. In other cases it may be appropriate to have a system of further payments related to increased capacity, for example, where a paid-up royalty or its equivalent has been associated with the Contract. In such cases a Special Condition should be written to deal with this and other departures from Clause 8 to suit the different circumstances which may apply.

The obligations of the Contractor in relation to information supplied by the Purchaser, as set out in Clause 20, open the way for special measures to ensure confidentiality. This may be particularly important when the Contractor carries out the detailed design or construction of a Plant using the Purchaser's own process. In such cases it is sometimes desired to restrict the circulation of Documentation and to ask for return to the Purchaser of all copies of Documentation at the end of the Contract, or to have other security measures of this kind. Opinions differ on their practical effectiveness but if some deterrent of this kind is required for the Purchaser's protection a suitable Special Condition should be prepared.

The obligations of the Contract become effective only after the Purchaser and the Contractor have entered into a valid contract. As indicated above if any confidential information is communicated by either party at the tender stage it should be protected by suitable undertakings given at that stage.

Suitable wording for a Special Condition might be:

**EXAMPLE CLAUSE**

## 20. Confidentiality

20.6 If the parties have entered into a separate agreement to govern **Confidential Information**, such agreement shall govern and take precedence over the provisions in the **Contract** relating to the disclosure of **Confidential Information**.

---

### Intellectual property rights (IPRs)

Considerable care must be taken when the Contract is being drafted to ensure that all interested parties that have any intellectual property rights are properly protected. The possible need for Special Conditions in this regard is extremely important, particularly where the Contractor is licensing a process or where a Subcontractor or supplier is responsible for the provision of specialised equipment and know-how. The Purchaser and the Contractor must take every reasonable precaution to ensure that their respective specialist suppliers and Subcontractors are properly protected under the Contract and that both parties are aware of any implications in the disclosure or use of information being provided by one to the other.

The clear vesting in Clause 8 of copyright with the Contractor and the Purchaser being granted a licence to use the Documentation has overcome the need to draft Special Conditions clarifying the rights of both parties, though it may be desirable to make provision for copyright to pass to the Purchaser in the event of termination for default.

Sub-clause 8.3 limits the liability of the Contractor for any infringement of any intellectual property right protected by law to those existing at the date of tender. There always exists a small possibility of a patent being granted to a third party in the period between this date and that of the Plant coming into operation. The effect of Sub-clause 8.3 is that in this case any cost of altering the Plant to avoid infringement or the payment of royalties would be borne by the Purchaser. If the Purchaser is concerned about this aspect of the Contract then the use of a Special Condition should be discussed with the Contractor.

If the process is supplied by the Contractor as part of his expertise then it would be reasonable to expect him to review patents in the field and inform the Purchaser of any that might be relevant to the design, construction and operation of the Plant and any involving product use.

### Images of the Plant

Sub-clause 20.4 clarifies the question of when the Contractor has the right to take photographs or make any other record of the Plant, whether on film, video recording or some other form of image record. As video and electronic photography are often a necessary tool for communication between the construction site and the design office on a regular or continuous basis this must be specifically permitted under the Contract. The Purchaser retains the right to decide whether images can be used for publicity purposes.

### Technology development

A further important area is that of new inventions or developments created in carrying out the Works. It may be the case that the Works will include the development of new processes or products, and each of the parties may see itself as the rightful owner of the associated intellectual property rights and entitled to patent or otherwise protect the new technology for its own commercial benefit.

Who is to own the new invention or development? Is it the Purchaser, who is funding the development, or the Contractor, whose expertise has mostly produced the development? Another possibility is that the development work may have been undertaken by a specialist Subcontractor or licensor. Where the parties are working closely together they may each have an equally strong argument for owning the rights, and joint ownership may be the answer although that will have longer term commercial consequences.

With various possible solutions available, the choice will depend upon the particular circumstances both as to the identity, policies and expertise of the parties, the particular technology and the degree of collaborative working. The General Conditions are mostly silent on this important matter. The only provisions in the General Conditions which are relevant are Sub-clauses 8.7 and 8.8 which deal with copyright in Documentation provided by the Contractor or his Subcontractors for which the Purchaser is granted a right to use such Documentation in connection with the Plant, Sub-clause 8.9 (see below), and Clause 20 which is a typical confidentiality provision.

Where the Works are likely to include the development of new technology the parties will need to establish and agree their approach and a suitable Special Condition will need to be included in the Contract. In doing so it may be necessary to amend Sub-clauses 8.7 and 8.8. If long-term licences are envisaged, other than for the development and operation of the Plant, it may be appropriate for the parties to enter into a separate licensing agreement.

### Software

Note the Contractor's obligation in Sub-clause 8.9 regarding the software that he provides. Any further requirement, e.g. source codes, should be detailed in a Special Condition or in Schedule 2.

## Guide Note C: Unforeseen physical conditions

*See Clause 6 (Sufficiency of Contract Price) and Schedule 18 (Valuation of Variations and claims)*

For a variety of different reasons the Contractor may not have been able to inspect the Site adequately before submitting his tender and agreeing the Contract and Contract Price. There is also the problem of dealing with the consequences of any unforeseen or unforeseeable conditions found at the Site during the construction phase.

Unless a full detailed survey of the proposed Site has been carried out by, or for, the Purchaser at the start of the bidding process, and the Contractor has been able to take account of that survey in compiling his tender it will normally be necessary to survey the Site after the Contract has been signed. This will enable at least some of the major potential difficulties with ground conditions, etc, at the Site to be identified as early as possible during the Contract, so that the Contractor can allow for them.

Site surveys undertaken after the Contract has been made will usually be carried out by either the Contractor alone or possibly by the Purchaser and the Contractor acting together. It can be the case that the Purchaser alone will be responsible for Site surveys, but this is comparatively rare. It will be necessary to allow for these surveys in the Contract, especially with regard to Contract Price, programme, and the machinery for dealing with any problems brought to light by the surveys.

It such circumstances the Contractor should state in his tender the assumptions as to the conditions at the Site upon which his tender and price(s) are based. The Contract will need to include a Special Condition to cover the intended survey. In addition, Sub-clause 6.3 will require amendment so that following the issue of a Variation Order to adjust the Contract Price in the light of the findings of the Site survey (see example Sub-clause 6.4 below) the date of that Variation Order becomes the reference date for considering the Contractor's knowledge of the Site in any subsequent application of Sub-clause 6.3. Example clauses are as follows:

---

**EXAMPLE CLAUSE**

## 6. Sufficiency of Contract Price

Delete the text of Sub-clause 6.3 in the General Conditions and insert:

6.3 If during the carrying out of the **Works** the **Contractor** encounters on the **Site** any physical condition which at the date of the **Contractor's** tender as stated in the **Agreement** or, following the issue of a **Variation Order** in accordance with Sub-clause 6.4, at the date of such **Variation Order** could not reasonably have been foreseen by a properly qualified and competent contractor having all the information which the **Contractor** then had or could have obtained by a visual inspection of the **Site** or by reasonable enquiry, and if the **Contractor** considers that he will in consequence of such condition incur an increase in the time or **Cost** of performing his obligations under the **Contract**, he shall notify the **Project Manager** within fourteen days of becoming aware of such unforeseen condition and otherwise shall comply with the requirements of Clause 19 (Claims). Any such notice shall:

(a) expressly state that it is given under this Sub-clause 6.3;
(b) specify the condition encountered;
(c) specify the steps which the **Contractor** is taking or proposing to take to overcome the condition encountered;
(d) contain estimates of the effect on the **Approved Programme** of the condition and of the additional **Cost** which the **Contractor** is likely to incur; and
(e) state how the **Contractor** proposes to minimise such effect and additional **Cost** and/or time.

If the **Contractor** is unable to include in such notification any of the information listed in (b) to (e) above, he shall provide it to the **Project Manager** in accordance with Clause 19.

The **Contractor** shall be entitled to an addition to the **Contract Price** determined in accordance with and subject to Clause 19 in respect of the additional reasonably incurred **Cost** of:

(i) complying with any instruction of the **Project Manager**; or
(ii) in the absence of any instruction from the **Project Manager**, doing whatever is reasonably necessary;

in consequence of the conditions specified in a notification issued under this Sub-clause 6.3.

6.4 The **Contractor** shall carry out within the period stated in **Schedule 11** (Times of completion) whatever geotechnical and other surveys of the **Site** are appropriate, and deliver a copy of the report(s) resulting from those surveys to both the **Purchaser** and the **Project Manager**. In the event that the reports show that the conditions at the **Site** are significantly different to those allowed for by the **Contractor**, as stated in the tender, the **Contractor** may give notice to the **Project Manager** in accordance with Clause 6.3 within fourteen days after delivery of the said reports, and the **Project Manager** shall issue a **Variation Order** to the **Contractor** stating an appropriate increase or decrease as the case may be to the **Contract Price** in accordance with the provisions set out in **Schedule 18** (Valuation of Variations and claims).

---

The parties will then need to allow for Site surveys in Schedule 11 (Times of completion), and may also wish to make special provision in Schedule 18 (Valuation of Variations and claims).

It is a matter of fact whether particular Site conditions are unforeseen, to be determined by the Project Manager. Factors to be taken into account will include the availability of published information on the local ground conditions, what relevant conditions might be inferred from a visual examination of the Site and the extent and reliability of Site investigation data made available.

It may be that the Purchaser wishes the Contractor to be wholly responsible for the physical conditions at the Site with no opportunity for the Contractor to claim extra money or time under Sub-clause 6.3. The Purchaser must be aware that in such circumstances the Contractor will have to cover his additional risk by a higher Contract Price. For this reason this approach is not recommended, although it is recognised that in some circumstances (for example at the insistence of any lenders) this approach is unavoidable. The necessary changes to the General Conditions would be the deletion of Sub-clause 6.3 and, for the avoidance of doubt, the insertion of a new Sub-clause which states that the Contractor accepts total responsibility for having foreseen all difficulties and costs of carrying out his obligations and that the Contract Price will be not be adjusted nor any extension of time given to take account of any unforeseen conditions.

It is always the case that the earlier that problems are identified the cheaper the solution in the end. For that reason it is always preferable to have a Site investigation survey carried out before tenders are invited, with all information from the survey being available to tenderers.

## Guide Note D: Taxes

The primary intention of Clause 7.4 is to ensure that the Contractor prices and makes allowance for all corporate and personal taxation associated with the performance of the Contract. Note that Clause 41.13 makes the Contract Price exclusive of Value Added Tax.

## Guide Note E: Site services and working conditions

*See Clauses 12 (Contract Manager and Contractor's staff), 26 (Health, safety, environment and pollution), 27 (Site services) and 28 (Site working conditions)*

Clauses 12, 26, 27 and 28 deal with the Site as regards labour, working conditions and all necessary facilities in general. The details should be set out in Schedule 1 (Description of the Works) or Schedule 3 (Responsibilities of Purchaser) bearing in mind the following:

(a) Labour and supervision: usually provided by the Contractor;
(b) Housing and transport, canteen or other catering facilities for construction labour: the extent to which they are provided, by whom and their location;
(c) Sanitary, first-aid and other medical facilities: always required, but may be provided by either party;
(d) Contractor's Equipment, unloading facilities, storage huts or compounds, security arrangements, and the like: usually provided by the Contractor;
(e) Fire-fighting and other safety measures: local circumstances vary widely;
(f) Water, electricity and other utilities for construction purposes: usual for the Purchaser to arrange for supply to a point on the Site and for the Contractor to provide the necessary distribution. Consider the cost and payment (if any), availability (time and capacity) and temporary arrangements;
(g) Consumable materials, utilities or facilities used for the carrying out of take-over procedures: usual for the supplies provided for the Contractor by the Purchaser to continue until the issue of the Final Certificate;
(h) Temporary office facilities including telephones, fax, email and other communications;
(i) Site working conditions: rates of pay and working conditions may be subject to any special site agreement between employers and trade unions; productivity bonus schemes of various kinds may be stipulated either as part of such site agreements or separately;
(j) Site security including access control;
(k) Laydown areas, pre-fabrication facilities and their location;
(l) Site safety programs, plans and training.

Clause 26 sets out the requirements for compliance with all health, safety and environmental legislation. The details of the applicable HS&E requirements should be included in Schedule 4 (Health & Safety) and 5 (Environmental protection & waste disposal).

Before construction commences, clear and concise site rules should be drafted and all necessary safety documentation and equipment be made available to ensure safe working conditions at the site. The Contractor should be proactive in establishing all necessary safety plans but should expect the full support of the Purchaser in the implementation of safety throughout the project.

The Contractor should define his requirements for lay down areas and pre-fabrication facilities so that sufficient space is provided by the Purchaser.

## Guide Note F: Variations and claims

*See Clauses 16 (Variations), 17 (Variations proposed by Contractor), 18 (Valuation of Variations) and 19 (Claims) and Schedule 18 (Valuation of Variations and claims)*

The use of a lump sum contract dictates that at the date of contract award the work to be done by the Contractor is well defined and accurately priced.

Nevertheless it is usual to make provision in lump sum contracts for the possibility of changes in the type and extent of the work to be done by the Contractor in respect of the Works. The need for changes can arise for a variety of reasons; perhaps because the Purchaser's requirements are modified or because the Contractor can recommend improvements from recent engineering experience. Note that the cumulative effect of too many variations can be detrimental to the contract.

Careful and prompt use of the Variation procedure to record detailed changes of the Specification and Description of the Works is essential for the good management of the project and for the avoidance of disputes. Full discussion and cooperation between the parties during the Variation procedure also avoids disputes.

Clause 16 establishes the Project Manager's right to require a Variation, either with or without first obtaining a quotation or agreeing all the contractual implications, while protecting the Contractor from major disruption of his forward business or from having to accept unreasonable changes in his responsibilities. The figures of 25 per cent and 5 per cent specified in Sub-clause 16.6(a) and (b) may be changed if required by a Special Condition.

Clause 17 allows for Variations proposed by the Contractor. The Project Manager is required to consider such proposals and inform the Contractor whether or not he intends to order such a Variation. The Project Manager's decision is final except where a rejected proposal is intended to eliminate a potential Defect or hazard. However, if the Variation relates to a matter which should have been taken into consideration by the Contractor in preparing his price for the Works or if the need for the Variation arose as a consequence of a failure of the Contractor, the Contractor will not be paid the cost of implementing the Variation.

Clause 18 establishes the basis for adjustments to the Contract Price as a result of a Variation being incorporated into the Works. In cases where the Project Manager and the Contractor are not able to agree on the change to be made to the Contract Price, the matter must be decided by the Project Manager. Sub-clause 16.3 allows for the Project Manager requiring the Contractor to prepare an estimate of each Variation in such detail as the Project Manager requires. If for reasons outside his control the Contractor is unable to give an accurate estimate, he should at least provide an order of magnitude cost for the Variation which will establish a basis for negotiation when more detailed costing information is available. The Contractor is also required to maintain adequate records of the cost of doing the extra work.

Clause 19 sets out the procedures for the submission of claims by the Contractor and their consideration by the Project Manager.

Any departure from the Specification may have complex consequences not immediately apparent, which may have cost and time implications. There have been cases in which severe financial strains have been imposed on either the Purchaser or the Contractor by a series of Variations that have been costed either too high or too low. Experience shows that it is wise to prepare a properly costed Variation, as a quick approximation usually results in too low an estimate.

The limited scope of application of Variations should be noted. Any amendments to the Contract which are not alterations to the Plant, or to the Approved Programme or the type or extent of the Works (for example, changes to any of the Schedules other than Schedule 1 (Description of the Works)) must be made by written agreement. This could simply consist of an exchange of letters.

## Guide Note G: Delivery to Site

*See Clause 24 (Delivery to Site)*

The General Conditions are written on the basis that the Contractor will be responsible for arranging the delivery of the Materials to the Site. In respect of any material to be provided by the Purchaser the delivery arrangements should be set out in Schedule 3 (Responsibilities of Purchaser) (Sub-clause 24.1).

## Guide Note H: Ownership

*See Clause 25 (Ownership of Materials)*

Clause 25 defines the arrangements as to when ownership or title in the Materials which will form part of the Plant passes to the Purchaser. This will usually be on payment being made to the Contractor or delivery to Site, whichever is the earlier.

As vesting certificates have on occasions proved to be worthless, the Purchaser would be well advised to ensure that the Project Manager exercises his authority under Sub-clause 25.3. This is particularly so when a stage payment to a Subcontractor, made under the terms of a subcontract for a major element of Materials, is to be reimbursed to the Contractor by the Purchaser in respect of Materials which have yet to be delivered to the Site. In such circumstances Sub-clause 25.2 is also applicable.

## Guide Note I: Insurance

See *Clause 30 (Care of the Works) and 31 (Insurance)*

Clause 30 (Care of the Works) sets out the obligations and liabilities of the parties with respect to the care of and damage to the Materials, the Temporary Works and the Plant, and also regarding injury to persons and property, all in connection with the performance of the Works and provision of the Plant. Given the potentially high value of these losses it is recommended that insurance is placed to transfer this risk to an insurer.

Clause 31 (Insurance) sets out the obligations of the parties to procure insurance for personal injury and damage to property associated with the project or proximate to the Site. It is inevitable that the two clauses, though covering different ground, are considered together as their topics are very inter-related.

It is generally in the best interests of both the Purchaser and the Contractor to arrange matters so that one and only one insurer is responsible for any one area of risk. The aim is to avoid double insurance where the same risk is covered by two different policies of insurance which inevitably leads to unnecessary expense and potentially to conflict and counterclaims between the two insurance companies involved. This means that a single insurance policy should cover the specific damage regardless of who is at fault in causing the loss. The General Conditions allocate insurance responsibilities to avoid such duplication, and usually stipulate that the party responsible for placing the insurance is ultimately liable for any exclusion in the policy or shortfall in actual recovery from the insurer.

#### Materials until delivery to the Site

Sub-clauses 24.5 and 24.6 require the Contractor to be responsible for all Materials until delivery to the Site including any free-issue material that the Contractor first takes possession of at a location other than the Site. Sub-clause 31.2 requires the Contractor to procure, or arrange for the Subcontractors to procure, all risk insurance to cover the Materials until delivery to the Site. This requirement is extended to free-issue materials if the Contractor first takes possession of them elsewhere than at the Site.

#### Plant, Temporary Works, Site Materials and Documentation

Sub-clause 31.1 requires the Purchaser to procure the Construction All Risk (CAR) insurance (also known as Contractor's All Risk or Builder's All Risk insurance) to cover the Plant, Temporary Works and Site Materials. The CAR insurance policy shall include cover during the Defects Liability Period on the equivalent of "Guarantee maintenance" basis covering any cause prior to Take Over of the Plant and any activity of the Contractor performed in connection with the Contract during the Defects Liability Period. The CAR insurance policy is also to include for defective design, specification or workmanship in accordance with The London Engineering Group Model "Consequences" Defects wording (LEG2/96). This approach covers the damage to the property except for the cost that would have been incurred to rectify the defect had it been rectified immediately prior to the occurrence of the damage.

Sub-clause 30.4 requires the Contractor to make good loss and damage to Site Materials, Temporary Works and the Plant prior to Take Over of the Plant by the Purchaser and to a limited extent thereafter. The Contractor is liable for only two specific costs incurred in making good the damage. Firstly, the Contractor is liable for an initial amount of loss (the defined Deductible as agreed between the parties) unless the loss or damage is due to a Purchaser's Risk (Sub-clauses 30.5 and 30.6). Secondly, the Contractor retains liability for Defects under Clause 37 (Liability for Defects) to the extent of the cost to rectify the defective design, specification or workmanship had it been rectified immediately prior to the occurrence of the damage. This provision therefore covers the costs excluded under the CAR insurance policy on the basis of LEG2/96 defects wording. Otherwise the Purchaser is responsible for providing the funds to make good the damage to Site Materials, Temporary Works and the Plant either from the insurance proceeds or the Purchaser's own funds (Sub-clauses 30.7 and 30.8).

Sub-clause 30.9 requires the Contractor to make good loss and damage to Documentation prior to Take Over of the Plant.

Note that any delay in issuing the Final Certificate will have an effect on the duration of the Purchaser's obligation under Sub-clause 31.1 to provide CAR insurance. See also the guidance on compiling Schedule 16 (Performance tests and procedures) and Guide Note L (Liability for Defects).

The Purchaser may wish to transfer to the Contractor the risk of compliance with certain conditions and warranties contained in the CAR policy the breach of which may result in the failure of the policy to respond either in part or whole to a claim. This may include requirements concerning the availability at site of specific fire fighting facilities or restrictions on the construction methodology such as limits on the amount of open trench at any time. Any such transfer should be specifically discussed and agreed by the parties in conjunction with their professional advisers to provide the detailed drafting of an appropriate Special Condition.

#### Loss or damage to the other property of the Purchaser and his Affiliates

Sub-clause 30.13 states that the Contractor is responsible for any damage to the property of the Purchaser and his Affiliates up to an amount stated in the Agreement, or in the absence of such an amount up to £5 million. The associated insurance is to be provided as part of the insurance required of the Contractor in accordance with Sub-clause 31.4. The Purchaser takes responsibility through an indemnity to the Contractor for losses in excess of the amount stated in Sub-clause 30.13 (Sub-clause 30.14). The Contractor should pass his liability under Sub-clause 30.13 onto Subcontractors in any relevant subcontract.

#### Loss or damage to the other property of the Contractor and his Subcontractors

The Contractor is responsible for any damage to his own other property or that of his Subcontractors even where the loss is due to the negligence of the Purchaser (Sub-clause 30.12). Sub-clause 31.3 only places an express requirement on the Contractor to procure insurance for property brought to Site by the Contractor or his Subcontractors (defined as Contractor's Equipment). The Contractor should nonetheless insure all of his property in accordance with his standard risk management policies. The Contractor should also pass an equivalent liability onto Subcontractors in any relevant subcontract.

#### Death or injury to the employees of the Purchaser (and Affiliates) and the Contractor (and Subcontractors)

The intention behind the Contract is that each party involved in the Works should be made responsible for injury or death to his own employees even where due to the fault of the other party (Sub-clauses 30.10 and 30.11). This position is extended to cover the employees of Affiliates and Subcontractors. This apportionment of risk on a "knock-for-knock" basis is commonly used on construction sites in the United Kingdom. This arrangement will need to be endorsed by all relevant insurers. Provisions similar to Sub-clauses 30.10 and 30.11 must therefore be included in each subcontract or consideration given to a site-wide reciprocal hold harmless and indemnity agreement. Sub-clause 31.5 sets in the Contract the statutory requirement on the parties for Employer's Liability insurance.

#### Loss of or damage to third party property

Sub-clause 31.4 requires the Contractor to procure third party liability insurance to cover loss of or damage to third party property and injury to third party personnel caused by the fault of the Contractor or any Subcontractor in the sum of at least £10 million. Under Sub-clause 30.15, the Contractor is required to indemnify the Purchaser against any third party claims to the extent caused by the fault of the Contractor or any Subcontractor. The Contractor should pass this liability onto Subcontractors in any relevant subcontract.

The parties should carefully consider any risk arising from the specific circumstance of the project, in conjunction with their professional advisers, to ensure that appropriate coverage is in place. For example, particular care needs to be taken where high value facilities not owned by the Purchaser or his Affiliates are adjacent to the Site.

#### General

As a default position the Contractor is not required under the contract to provide any professional indemnity insurance (Sub-clause 31.6). The parties should consider with their professional advisers whether such insurance should be provided for their particular contract.

The parties should also note the indemnities in Sub-clause 26.7, which set out their respective responsibilities for clean-up costs in the event of a pollution incident. The final paragraph of Sub-clause 26.7 is based on the distinction between damage to property caused by pollution (dealt with under Clause 30) and the clean up costs of pollution (Sub-clause 26.7).

Sub-clause 31.7 covers the right of each party to obtain details of the insurance cover of the other. In the context of drafting a published form of contract it is not possible to be more specific. Within a particular contract the parties may wish to include one or more Special Conditions, setting out their requirements in more detail.

If the parties wish to modify any of the provisions of the Contract this should be done by passing any revised requirements to their professional advisers for the detailed drafting of Special Conditions. In any event, specialist advice should be sought as to the appropriateness of the care of the Works and insurance provisions for the particular project. Specifically consideration should be given to ensuring that all those whose interests may be affected by the project are appropriately taken into account in the insurance and indemnity arrangement. This may require a Special Condition to cover the involvement of such parties such as process licensors, Purchaser's co-venturers, occupier's of adjacent property, operators, government bodies etc. For particularly complex arrangements the parties should consider a specific project-wide mutual indemnity and waiver of recourse agreement.

## Guide Note J: Completion, Taking Over, testing and start up

#### Completion of construction and Taking Over

The concept of completion by the Contractor of the Works has several significant aspects:

(a) compliance with the Specification and with the Description of the Works and satisfactory completion of tests and procedures;
(b) timely finishing of particular tasks in accordance with the Approved Programme with reference as appropriate to financial incentives;
(c) transfer of responsibility.

The General Conditions have been devised to cater for each of these by both procedure and documentation. A summary is given in Table 2.

**Table 2**—Procedures and documentation for completion of construction, Take Over, Acceptance and issue of Final Certificate

| Topic | Clause or Sub-clause of General Conditions | Document defining contractual requirement | Record of completion | Certificate prepared by |
|---|---|---|---|---|
| The Plant or any part of it | 32.2<br>32.3<br>32.6 | Specification:<br>Schedule 14 (Criteria for the completion of construction) | Construction Completion Certificate | Contractor (draft); Project Manager may endorse. Project Manager to finalise |
| Time | 32.8 | Schedule 11 (Times of completion) | Construction Completion Certificate | Project Manager |
| Responsibility for the Plant | 33.7 | Schedule 15 (Take-over procedures) | Take Over Certificate | Project Manager |
| Plant performance | 35<br>36 | Schedule 16 (Performance tests and procedures) Schedule 17 (Performance guarantees and damages for failure) | Acceptance Certificate | Project Manager |
| Final completion and end of final defects liability period | 38 | Specification; Schedule 1 (Description of the Works) | Final Certificate | Project Manager |

All certificates are issued by the Project Manager to the Purchaser and the Contractor. An example of each certificate is included at Appendix B.

Completion of construction should only occur when everything described in the Specification is physically complete, and when any prescribed tests have been successful (see Schedule 14 (Criteria for the completion of construction)).

The terms 'start-up', 'pre-commissioning', 'mechanical completion' and 'commissioning' have not been used in the General Conditions. They are not sufficiently well defined to enable the Project Manager to issue a certificate with a precise date. To concentrate on any single discipline would be misleading.

It is recommended that Completion of Construction is used in Schedule 11 (Times of completion) as a point at which any delay results in payment of liquidated damages, if they are to be applied. The criteria which are to be satisfied by the Contractor before the issue of a certificate are defined in Schedule 14. However, if the Purchaser wishes to have liquidated damages applied to delay in Taking Over, the procedure to be followed is described at the end of the guidance on compiling Schedule 11.

Take Over is another important event. It is very important from both contractual and safety points of view that responsibility for the Plant should pass from the Contractor to the Purchaser in a clearly defined and unambiguous manner (see Sub-clause 33.7). The time for this to be done could be just before any raw material is first fed into the Plant; however, in some circumstances the Contractor may take responsibility for the initial operation of the plant during a take over test. From the time of Take Over the Purchaser is responsible for the care, operation, maintenance and safety of the Plant. It should also be noted that the Defects Liability Period starts on the date of Taking Over.

Except in the simplest of projects, the Plant is likely to consist of several Sections, some of which are operationally dependent on others. There will thus be a logical sequence in which completion of each Section will be required in order to achieve satisfactory commissioning of the whole Plant. Furthermore, the actual task of thoroughly inspecting and testing a completed Plant before it can be used safely can be costly in time and resources. To leave the whole of a project to be inspected and tested at the same time could lead to the procedure being rushed because of pressure for production to start and hence lead to potential hazards, or else to delay if the inspection work is to be done properly. For these reasons the Approved Programme should specify the sequence of completing the different Sections in some detail, and should allow adequate time for inspection and testing. The Sections should be described in the Specification and then referred to in Schedules 11 and 12.

Sub-clauses 32.1, 33.1 and 36.2 provide respectively for completion of construction, Taking Over and Acceptance to be applied separately to each Section of the Plant.

For each Section of the Plant the following list, although not complete or appropriate for all projects in every respect, may be used as an aide-memoire in identifying those activities which should be included in Schedule 14 as part of construction (possibly up to and including item (l)), and those activities which should be classed as take-over procedures (possibly (m) to (t)). The remainder could then be carried out by the Purchaser and his staff after Taking Over with advice from the Contractor, who should be able to observe the actions of the Purchaser sufficiently closely to satisfy himself that his operating and maintenance instructions, where applicable, are being observed:

(a) Handing over of Test Certificates, in particular those required by statute or insurers, relating to any item of Materials;

(b) Installation of gland packing and lubrication of valves and minor machinery; checking correct rotation of drivers;
(c) Isolating equipment, removing expansion joints, rupture discs and relief valves from pipework and equipment before line flushing, cleaning and testing and reinstalling thereafter;
(d) Removing loose material and dirt, installation of internals, inspections and closing up of vessels;
(e) Cleaning pipework by flushing, steaming, blowing, pigging or scatter scaling;
(f) Hydraulic and/or pneumatic pressure testing of equipment and pipework;
(g) Applying special treatment or other preparation of internal surfaces;
(h) Setting and testing non-return valves, relief valves and other relief systems e.g. bursting discs;
(i) Installing orifice plates after line cleaning, installation of temporary strainers, removal of slip plates;
(j) Checking and testing electrical installations;
(k) Checking alignment of major machinery and drivers; cleaning and flushing lubricating oil installations; checking direction of rotation of motors and drivers;
(l) Checking and adjusting safety systems;
(m) Conducting short running-in tests on machinery;
(n) Checking action of instruments and control valves, testing electrical controls, alarms, computer systems and software;
(o) Calibrating instruments;
(p) Drying out or purging plant;
(q) Drying out refractories;
(r) Adjusting pipe supports for expansion;
(s) Removing, cleaning and replacing temporary strainers;
(t) Checking orifice plates and permanent blanks, relief systems, drainage systems, safety provisions;
(u) Charging catalysts;
(v) Charging raw materials, process chemicals, fuel;
(w) Warming up, starting up fluid flows;
(x) Tightening hot joints;
(y) Starting up and operating the various Sections of the Plant;
(z) Routine maintenance, cleaning, plant adjustments.

At this stage final safety reviews prior to start up should be carried out.

**Some of these actions may reveal construction errors, which will then be rectified while the other activities mentioned are proceeding. Insulation and painting work is also likely to go on at this time, together with minor construction details not previously finished.**

### Plant start-up and performance tests

For the majority of process plant projects the later phases of construction, the preparation for start-up and the early phases of operation proceed more or less continuously without clear boundaries. The work done in this transition period is customarily called 'commissioning', but as this is an ill-defined term it is not used in the General Conditions. Nevertheless commissioning is an essential part of the work required to provide a completed process plant which operates in accordance with the designer's intentions. It is usual for the first attempt to start up a Plant to reveal weaknesses or omissions, however small. It is therefore customary to plan the start up in stages, so that if a shutdown or pause is required for rectification or adjustment, this can be done in an orderly manner at the minimum cost of time and resources. To be successful a Plant must not only be complete and tested but it must also have feedstocks, intermediate chemicals and utilities and must be served by properly trained and competent operating and maintenance teams. The organisation of the commissioning period within an agreed procedure is very important. Responsibilities, particularly those relating to the person in charge of commissioning, must be clearly defined and lines of communication must be set up.

The advantage of the Purchaser's staff having the responsibility of operating and maintaining the Plant from start up is that it forms a culmination of their training and makes them concentrate on following meticulously the Contractor's instructions and the procedures specified in the operation and maintenance manuals, operator's guides, etc, which will incorporate information provided by Subcontractors and suppliers. If they fail to follow these instructions and the Contractor can show such failure, then he may make out a case to be excused from achieving the guaranteed results in a particular test or he may be able to claim that a defect which appears is not his responsibility.

Sub-clause 35.5 provides that the performance tests are to be carried out by the Purchaser in the presence of the Contractor, with the operating and maintenance staff needed to start up and operate the Plant being provided by the Purchaser. It may be appropriate also for the licensor to be present during the performance tests. This has generally been found to be the best arrangement when the process technology and performance guarantees are provided by the Contractor or licensor. As a rule the personnel required will have been selected and trained during the period of procurement and construction of the Plant and arrangements will have been made for specialised training by the Contractor or by equipment suppliers or licensors, in which case such arrangements should be described in Schedule 9 (Training by Contractor). The Contractor's staff who are present during the Plant start-up support the Purchaser's staff and assist in decision-making.

It is in the interests of both parties for the Contractor to observe all operations closely until the performance tests have been successfully completed so that the Plant is safeguarded against damage by faulty operation which, if caused by the negligence of the Purchaser's staff, may invalidate any guarantees given by the Contractor and supplier's warranties.

It is not unusual for the Contractor to provide for such experienced staff to remain on site at the request of the Purchaser, for as long as he may require, on a reimbursable basis, so that they can be released when the Purchaser's own staff are competent to proceed without them.

The content of Clause 35 is important. All those involved with the performance tests must understand it. The flowcharts provided at Appendix A (Performance Test Flowcharts) are intended to assist in that understanding.

### Minimum performance requirements stipulated in Schedule 17 (Performance guarantees and damages for failure)

Sub-clauses 35.12, 35.14 and 35.15 provide for the Purchaser to be compensated if the results of any final performance test are outside the guarantees specified in Schedule 17 for reasons which are the responsibility of the Contractor. See the detailed guidance on compiling Schedules 16 and 17.

## Guide Note K: Incentives for early completion

See *Clauses 13 (Times of completion and Approved Programme), 15 (Damages for delays) and 35 (Performance tests)*

Where it is particularly advantageous to the Purchaser to have the Plant in production or completed as early as possible, it may be appropriate to have an early completion bonus clause. The completion event(s) to which the bonus is related should be the same as those for liquidated damages for lateness.

As bonuses are the exception rather than the rule they are not included in the General Conditions. If a bonus for early completion is required, a Special Condition should therefore be written; it should state that the Contract Price would be increased by the amount of the bonus and give the method of calculation, the upper limit and any other conditions. Careful consideration should be given:

(a) to whether a bonus for timely completion should be linked to completion of construction or Take Over, with the latter being appropriate where the Contractor has substantial influence over, or even control of, all activities leading to Take Over, and

(b) to the effect of authorised extensions of time on the Contractor's entitlement to the bonus.

As with proposals for the payment of liquidated damages in the case of failure, any proposal by the Purchaser to pay bonuses should appear in the invitation to tender.

## Guide Note L: Liability for Defects

See *Clause 37 (Liability for Defects)*

Clause 37 concerns Defects in finished work or in Materials supplied by the Contractor and sets out the responsibilities and obligations of the parties when a Defect is discovered before or during the Defects Liability Period. The Defects Liability Period expires 12 months after the Taking Over of the Plant unless Sub-clause 37.5, 37.9 or 37.11 applies. The General Conditions provide that the Contractor is responsible for remedying the Defect. The Contractor will bear the cost of so doing if the Defect is the result of a breach of contract by him. In other circumstances, for example where a 'defect' is the result of a failure by the Purchaser correctly to operate the Plant, the Contractor's work is treated as a Variation and the Purchaser will bear the cost of the work.

Sub-clause 37.12 deals with the situation where the cost of repairing Defects exceeds any amount agreed in the Contract. If no limit is stated, the liability of the Contractor will be limited only by Sub-clause 45.3.

The provisions of Clause 37 regarding the length of the Defects Liability Period and its extension should only be modified if there are very good reasons for having different arrangements.

The guarantee period obtainable from the suppliers of major items of machinery or other equipment may sometimes expire well before the end of twelve months after Taking Over of the Plant, particularly if such items of equipment have been delivered by the makers to the Site a long time before the whole of the Plant is completed. This possibility must be taken into account by the Contractor during negotiations with the Purchaser on the terms of the Contract and the price. If the Contractor is not willing to carry the responsibility of remedying Defects in Subcontractors' work or Materials in accordance with Clause 37 he will have to obtain the Purchaser's agreement to a Special Condition providing for a suitable modification of Clause 37.

Certain items of equipment may be designed to have a working life which is short because the capital cost of an item with a long working life would be uneconomic. To avoid disputes about the working life of such items and the suitability of the designs, Schedule 10 (Parts with limited working life and spare parts) should list such items showing in 'running hours' the estimated working life and also any minimum life guaranteed under normal working conditions.

Defects may arise after the Defect Liability Period has expired. Sub-clause 38.3 makes it clear that the Contractor has no obligation to remedy such Defects. Equally, the Purchaser can make no claim against the Contractor in respect of Defects arising after the Defects Liability Period has expired.

Both Purchaser and Contractor should bear in mind that Clause 38 relates only to any liability which may arise between the parties to the Contract themselves. In addition, both the Purchaser and the Contractor may have other potential liability arising as a result of a Defect. That potential liability will be primarily of two types. The first will be criminal liability which may arise under laws relating to safety. The second type of potential liability involves civil liability for compensation to be paid for any loss or injury sustained by any person on the Site and other third parties as a result of the Defect.

Great caution is urged when amending Sub-clause 38.3. Nevertheless it is recognised that in matters of civil engineering works such a provision is very different from that industry's normal position, and it may be that in some situations where the Contract includes major civil engineering elements the Purchaser would wish that the Contractor is not relieved of his usual contractual obligations. If that is the case, a Special Condition should be drafted, with special care being taken to identify the scope of such a clause. Simple changes to Clause 37 regarding the length of the Defects Liability Period are not an appropriate way to deal with this issue. Equally a general relaxation of the limitation in Sub-clause 38.3, making for an increase in the Contractor's liability, is discouraged for the reasons stated above.

Any delay to the issue of a Final Certificate will have an effect where the Purchaser is providing insurance in accordance with Sub-clause 31.1, and the Purchaser should be aware of the continuing obligation to provide insurance that Sub-clause 31.1 imposes on him.

A Purchaser who wishes to benefit from an extended warranty covering one or more items of equipment, despite the general limitation imposed by Sub-clause 38.3, could introduce a Special Condition requiring the Contractor to provide an extended warranty to the Purchaser for those items of equipment. The Contractor will no doubt wish to secure a similar warranty from the appropriate Subcontractor. The Special Condition should also make clear that the extended warranty would apply notwithstanding Sub-clause 38.3. See also Guide Note R (Rights of third parties).

## Guide Note M: Suspension

*See Clause 42 (Suspension of the Works)*

There may be any number of reasons for the Project Manager to order the Contractor to suspend all or any part of the Works.

There may, for example, be outside influences beyond the control of either Project Manager or Purchaser, business or other reasons which warrant suspension by the Purchaser 'for convenience', or which result from the Contractor's actions or failures in some way. Since, except where he is at fault, the Contractor is entitled to recover additional cost incurred as the result of the suspension, it is clearly not in the Purchaser's interests to have the work suspended except where absolutely necessary; nor is it in his interests to allow suspension to be prolonged, particularly when the cause is within his control. Furthermore it must be appreciated that suspension involving interruptions to the supply of some types of major equipment may not be practicable. The Project Manager should always consider the impact on the time of completion, cost and any other aspect of the project before issuing a suspension order.

The Project Manager should note that Sub-clause 42.1 requires him to state the reason for the suspension. It is important that either at the time, or if that is not possible then later, the Project Manager states clearly if the suspension is because of the Contractor's default, as this will determine whether or not the Contractor is entitled to an extension of time and his Costs associated with the suspension.

Both the Project Manager and the Contractor must take notice of procedures and time constraints imposed by Sub-clause 42.5 in order to ensure that neither the Purchaser nor the Contractor is deprived of his rights.

It will be noted that Clause 42 does not incorporate any specific basis for the calculation of the sum due to the Contractor. There may or may not be an adequate basis set out in Schedule 19 (Contract Price and payment). The parties may wish to consider the inclusion of a Special Condition setting out in a comprehensive form the costs to which the Contractor will be entitled and the method by which they are to be quantified. In any event it is essential that full and detailed records are maintained throughout the period of suspension in order that any claim can be substantiated.

Sub-clause 42.3 requires the Project Manager to allow an extension to the project programme by a period of time not less than the duration of the suspension plus any time required by the Contractor to re-establish activities at the end of the period of suspension. If the Purchaser wishes the Contractor to reduce the extension period due to the suspension, it is likely that the Contractor will incur additional costs in accelerating his activities to reduce the extension. It could be appropriate for the Purchaser and the Contractor to agree upon a Variation under Clause 16 (Variations) to cover changes in working, costs or programme. The Contractor would then be entitled to such additional amount as may be reasonable including profit.

## Guide Note N: Termination

*See Clauses 43 (Termination by the Purchaser for convenience) and 44 (Termination for Contractor's default)*

There are two primary routes for contractual termination:

(a) under Clause 43, which provides the Purchaser with a contractual right to elect to terminate the Contract 'for convenience', irrespective of default by the Contractor;

(b) under Clause 44, which enables the Purchaser to terminate the employment of the Contractor in the event of the Contractor becoming insolvent as described in Sub-clause 44.1, or being in default as described in Sub-clause 44.2.

In the event of termination 'for convenience', Sub-clause 43.2 not only deals with the immediate cessation of the Works following receipt by the Contractor of the Notice of Termination (subject to any further work which the Project Manager requires – e.g. to make the Site safe), but also requires the Project Manager to issue one or more Take Over Certificates after which Clauses 37 (Liability for Defects) and 38 (Final Certificate) apply. Sub-clause 43.5 sets out the applicable payment provisions. It requires that the total amount due to the Contractor for the work done prior to termination together with any other costs the Contractor may be bound to pay (e.g. cancellation fees to suppliers) and any cost incurred by the Contractor in connection with the termination, such as in making the Site safe, be paid to the Contractor, subject to the payments already made.

In contrast, in the case of termination under Clause 44 Sub-clause 44.3 makes it clear that the Contractor is to forthwith leave the Site and thereafter has no right to return, except as the Project Manager may direct or permit. In the circumstances of termination for Contractor's default the calculation of the final amount due to the Contractor will be more complex. The Purchaser will normally engage another contractor to complete the Works, in which case the intention is that the original Contractor will be liable for any additional costs which are incurred by the Purchaser in having the Works completed and which are due to the Contractor's default. This is the approach taken in Sub-clause 44.6.

Sub-clause 44.7 sets out the payment arrangements if the Purchaser does not take steps to have the Works completed.

The payment arrangements in Clauses 43 and 44 must be considered in the context of the exclusive remedies clause (Sub-clause 45.2). In respect of termination for convenience under Clause 43, Sub-clause 45.2 limits the Contractor's entitlement to payment to those matters expressly stated in Clause 43. However, in the case of termination of the Contractor's employment under Clause 44 or a repudiation of the Contract by either party, Sub-clause 45.2 does not apply and the Contractor would be at risk of the Purchaser making further claims in law for breach of contract. See also Guide Note O (Limitation of Contractor's liability) for a more general description of the exclusive remedies clause.

Irrespective of the contractual termination route taken, Sub-clause 43.9 or 44.13 as applicable identifies those provisions of the Contract which survive the termination.

In addition, Sub-clause 14.9 provides a right of termination of the employment of the Contractor exercisable by either party in the event of protracted continuance of force majeure circumstances. Furthermore, the Contractor may in certain circumstances terminate his employment under the Contract where there is persistent failure on the part of the Purchaser to make due payments (Sub-clause 41.10) or where there is prolonged suspension of the Contractor's performance of his obligations by order of the Project Manager (Sub-clause 42.5).

## Guide Note O: Limitation of Contractor's liability

*See Clause 45 (Limitations on liabilities and remedies) and Guide Note L (Liability for Defects)*

Although rare in traditional building or civil engineering contracts, waivers or exclusions of liability for economic, indirect and consequential loss are invariably a feature of contracts in the process plant sector. The primary reason is one of scale; the Purchaser's capital investment, production costs and frequently turnover (particularly in the oil, gas, petrochemical and pharmaceutical sectors) are so high that it is neither reasonable nor practicable for the risk of such losses to be imposed upon the Contractor who is incapable of carrying it and, what is more, cannot normally insure himself against it. There is a general consensus among Purchasers and Contractors alike that it is impractical to expect the Contractor to carry all the risks of loss associated with design and construction in the process industry. This risk should therefore be taken into account by the Purchaser in making his original investment decision.

Clause 45 contains specific exclusions and limitations and there are provisions in the Agreement for maximum monetary limits to be placed on the Contractor's liability to the Purchaser for defaults under the Contract. However, users of the Conditions should note the last sentence of Sub-clause 45.2. The purpose of this proviso is to make it clear that in incorporating some limitations of the Contractor's liability there is no intention to cause the wording of the clause to be in conflict with the requirements of current legislation within the United Kingdom. The proviso makes it clear that the limit of liability stated in the Agreement does not in any way affect any liability of the Contractor to third parties under general law.

Sub-clause 45.2 is an 'exclusive remedies' clause. Except in the case of termination for Contractor's default under Clause 44 (Termination for Contractor's default) or repudiation by either party, Sub-clause 45.2 provides that only those remedies actually stated in the Contract are available to a party suffering a breach of the Contract by the other. In addition there are a number of specific exclusions and limitations of liability in Clause 45, and there are provisions in the Agreement for maximum monetary limits to be placed on the Contractor's liability to the Purchaser for defaults under the Contract. Note that it is only the 'exclusive remedies' provision in Sub-clause 45.2 which does not apply in the case of termination for Contractor's default; all the other provisions of Clause 45 apply in any event.

Attention is also drawn to Guide Note L and its commentary on Clause 38 (Final Certificate), which provides for a limitation of time on the Contractor's obligations.

## Guide Note P: Dispute resolution

*See Clauses 46 (Disputes), 47 (Adjudication), 48 (Reference to an Expert) and 49 (Arbitration)*

The Contract places a duty on the parties to use all reasonable endeavours to avoid disputes (Sub-clause 46.1) and, in the event a dispute does arise, to attempt in good faith to reach a settlement by negotiation (Sub-clause 46.6). This underlies a key part of the IChemE's contract philosophy that a project is best served by teamwork and co-operation between the parties, not by confrontation.

In the event that a disagreement cannot be resolved by negotiation, steps are set out (Sub-clauses 46.3 and 46.4) which must be followed by the parties if a dispute under the Contract is to be deemed to have arisen. With the exception of adjudication, short-cutting or ignoring these requirements, and a further series of requirements set out in the appropriate IChemE or other Rules (see below) may lead to legal problems later when a particular form of dispute resolution has been chosen.

The Contract provides (Sub-clause 46.7) for a dispute to be referred by agreement for mediation, for example to CEDR (the Centre for Effective Dispute Resolution, London), the CIArb (Chartered Institute of Arbitrators, London) or other similar bodies which publish appropriate rules for the conduct of mediation, if that is what the parties wish.

The General Conditions provide a variety of forms of dispute resolution to meet the needs of different types of dispute. The parties should consider carefully which form of dispute resolution is best suited to their particular needs. IChemE publishes Rules for the conduct of Expert Determination, of Arbitration, and of Adjudication, which are incorporated into the Contract as the default position. IChemE also publishes Rules for the conduct of Dispute Review Boards (DRBs) if the parties decide that a DRB is appropriate for their project.

But by far the best approach is to avoid disputes.

### Expert Determination

Recent experience in the process industries of the use of Expert Determination has been encouraging, particularly for single issue, essentially technical or valuation disputes. An Expert can bring his experience and professional knowledge directly to bear on resolving a dispute. Technical issues can prove extremely difficult for a legal arbitrator or arbitral tribunal, even when assisted by expert witnesses. Recently there have been increasing moves in other industries such as civil engineering to make use of Expert Determination. Properly applied, it is a relatively speedy and cost effective process for resolving complex technical disputes.

Expert Determination is normally carried out before a single Expert, with legal and dispute resolution experience as well as expert professional knowledge.

Unlike arbitration which is carried out within an established statutory framework, Expert Determination involves a contractual agreement between the parties and the appointed Expert for the Expert to decide the matter in dispute. It is, therefore, essential that the terms and conditions applying to all the parties are clear from the start. The Rules for Expert Determination, published by IChemE, are written with this requirement in mind.

A number of matters such as quantification of disputed Variation Orders are identified in the Contract as being matters which may be referred to an Expert for determination. Matters so identified may be referred by either party and do not require agreement between the parties to refer. Provision is also made for any other matter in dispute to be referred to an Expert for a final decision but only by agreement between the parties.

The powers of an Expert are set out in the IChemE's 'Rules for Expert Determination'. These also set out in detail the procedure to be followed by the parties in seeking the appointment of an Expert.

An Expert's findings are final, conclusive and binding on the parties (Sub-clause 48.7). Subject only to his either having failed to answer the questions put to him or having failed materially to act impartially between the parties, the Expert's findings cannot be appealed to the Courts.

Once a dispute has been determined by an Expert, it ceases to be referable to arbitration (Sub-clause 48.9). Parties should, therefore, be careful if there are a number of issues likely to need resolution not to arrive at a situation where some go to Expert Determination and some to arbitration, which would be undesirable because inconsistencies may result.

Expert Determination is particularly suited to those disputes where the knowledge and experience of a senior and respected practitioner in the field is likely to assist in reaching a speedy and fair result.

### Arbitration

Arbitration, although it has some similarities to Expert Determination, is carried out under the provisions of the Arbitration Act 1996*, as detailed in the IChemE's Arbitration Rules. The 1996 Act has moved a long way towards making English Arbitration awards final, by reducing substantially the number of matters that may be taken to the courts on appeal. Apart from 'partiality' by the arbitrator or his failing to answer the right questions, essentially the courts will only interfere on points of law and then only on points seen as substantially affecting the rights of the parties or as being of general public importance. Sub-clause 49.4 makes an arbitrator's award under this Contract final and binding upon the parties. If the parties wish to vary this requirement they should do so during Contract negotiations. To do so once a dispute has arisen will be extremely difficult.

The 1996 Act, in parallel with modern court procedure, places a duty on the arbitrator to 'manage' the reference efficiently and gives him substantial powers to enforce his duty, partly through the costs mechanism and partly through the power to disallow evidence submitted late. IChemE's Rules reinforce this position.

To cater for different types and magnitudes of disputes, IChemE has provided in its Rules for a variety of forms of arbitration (documents only, procedure for expert witnesses, etc.) as well as a full, formal procedure. The choice of the most appropriate form will be discussed by the arbitrator with the parties at the first meeting for directions.

If the parties do not wish to use IChemE's Arbitration Rules, they may by agreement use some other set of rules, such as CIMAR (Construction Industry Model Arbitration Rules). Any decision to change the 'default' rules in the Contract should be taken at the time of contract negotiation. Once a dispute has arisen such changes are very difficult to agree. The parties should also be aware that IChemE's Rules are consistent with IChemE Contracts and the Arbitration Act 1996. Choice of another set of Rules may lead to inconsistencies later.

Clause 49 constitutes a proper Arbitration Agreement as required by the 1996 Act. Changes to the wording of Sub-clauses 49.1 and 49.2 in particular should only be undertaken with legal advice.

### Adjudication

If a review of the Works and the Plant leads to the conclusion that the Contractor will be required under the Contract to undertake activities which are defined as construction operations by the Housing Grants, Construction and Regeneration Act 1996 Part II, (HGCRA) Clause 47 will apply to disputes arising from those activities, as will the IChemE Adjudication Rules.

If users prefer that adjudication should apply irrespective of the exemptions provided by the Act, this can be achieved by a Special Condition, which states that Sub-clause 47.1 is deleted.

It is strongly advised that users who believe that adjudication has no place in their contract do not attempt to delete the whole of Clause 47 by means of a Special Condition, for the simple reason that if a subsequent challenge establishes that the Act does apply then legislation will imply terms into the contract including the conduct of the adjudication under the Scheme for Construction Contracts and the selection of an adjudicator by any adjudicator nominating body irrespective of the knowledge of the form of contract, or, more importantly, process plant contracts.

Whilst it is hoped that parties will generally follow Sub-clauses 46.3 and 46.4, Sub-clause 47.2 makes it clear that for activities which do fall under the Act either party may call for adjudication at any time without first formally establishing a 'Dispute'.

Clause 47 has been modelled to correspond with Sections 108 and 108A of HGCRA, as amended by the Local Democracy, Economic Development and Construction Act 2009, under which an adjudicator's decision is binding until the dispute is finally determined by legal proceedings, by arbitration or by agreement. Sub-clause 47.8 of the Contract mirrors Section 108(3) of the Act.

While there is, as yet, no judicial authority on the point, it is arguable that adjudication under the Act can proceed even if an Expert Determination is also in hand. If so, it would appear that an Expert's finding given before an adjudicator has given his decision will stay the adjudication, while a finding given after a decision will over-ride the latter. But it would be far better to avoid the problem by choosing one approach or the other at the earliest possible stage.

### Dispute Review Board

Some major projects can benefit from the presence of a Dispute Review Board (DRB). A panel of specialists with experience in a variety of disciplines associated with the Contract Works is assembled at the commencement of the contract and meets regularly, perhaps every three months, through the life of the contract. The DRB receives briefings as to the progress of the contract, and if disputes arise they provide a first review forum. The DRB's decisions are not final, but they give the parties to the contract a good guide as to how the matters might be viewed in any arbitration or litigation. It is open to the parties to decide if they wish the findings of the DRB to be binding or non binding, a decision which should be taken at the time the discussions on setting up the DRB are taking place.

As a DRB will not be appropriate for all but the largest of projects, it is not appropriate to include a clause within the General Conditions to provide for a DRB. However, set out below is an example Clause 50 which may be adopted as a Special Condition as a means of providing for a DRB. Clause 50 is insufficient on its own: Clause 46 needs extensive amendment, as does Sub-clause 47.2. All of the necessary changes are set out below and can be incorporated into a Contract through Special Conditions.

IChemE's policy is for DRBs to make non binding decisions. However if the parties wish the DRB to make a binding decision (which can be reversed by agreement, by an Expert or by arbitration), the alternative version of Sub-clause 50.3 should be used.

The use of DRBs has become popular on major international projects, and indeed is mandatory on larger contracts funded by the World Bank. They are less common on domestic contracts other than the largest and most complex, not least because of the costs involved. If the parties do wish to appoint a DRB, this should be agreed in writing before the signing of the Agreement, and the General Conditions amended as described below. IChemE publishes Rules for the appointment and conduct of DRBs, which should be incorporated by reference in the documentation establishing the DRB in each case.

*In England, Wales or Northern Ireland. Similar provisions (whether by statute or in common law) apply elsewhere.

# 46. Disputes

Delete the text of Clause 46 in the General Conditions and insert:

**46.1** The **Purchaser** and the **Contractor** shall endeavour to avoid the escalation of problems into disputes as defined in Sub-clause 46.4 and to avoid disputes both between themselves and with third parties including **Subcontractors**. The parties acknowledge and agree that this shall not prejudice any right either party may have under Clause 47 to refer any dispute or difference to adjudication at any time, if that party so wishes, notwithstanding the definition of 'dispute' in Sub-clause 46.4.

**46.2** In order to avoid the development of disputes and to facilitate their clear definition and early resolution, the procedures set out in Clauses 46 (Disputes), 47 (Adjudication), 48 (Reference to an Expert) and 49 (Arbitration) shall be applied as appropriate.

**46.3** If the **Contractor** is dissatisfied with any **Decision** or valuation of the **Project Manager**, or of any person to whom the **Project Manager** may have delegated any of his authority or responsibility, or if the **Purchaser** or the **Contractor** is dissatisfied with any other matter arising under or in connection with the **Contract**, either party may at any time refer such dissatisfaction to the **Project Manager** giving full details of the nature of the matter. The **Project Manager** shall give a written **Decision** on the matter (giving the reasons for such **Decision**) to the **Purchaser** and the **Contractor** within twenty-eight days of such reference to him.

**46.4** The **Purchaser** and the **Contractor** agree that no matter shall constitute or give rise to a dispute unless it has been referred to the **Project Manager** under Sub-clause 46.3 and:

(a) the **Project Manager** has failed to give his **Decision** on the matter within the prescribed time; or

(b) a **Decision** given within the prescribed time is either unacceptable to the **Purchaser** and/or the **Contractor** or has not been implemented within twenty-one days of the **Decision**;

and, as a consequence, either the **Purchaser** or the **Contractor** has issued to the other a **Notice** (a '**Notice of Dispute**') setting out the nature of the dispute with a copy to the **Project Manager**.

The dispute shall be referred to the Dispute Review Board ('**DRB**') in accordance with Clause 50. The **DRB** shall give a written decision on the dispute (giving the reasons for such decision) to the **Purchaser**, the **Contractor** and the **Project Manager** within twenty-eight days of such reference to it or such other period as may be agreed between the parties in writing.

For the purposes of the performance of the **Works** and all matters arising out of or in connection with the **Contract**, the word 'dispute' shall be construed in accordance with this Sub-clause 46.4.

**46.5** Subject only to the provisions of Sub-clause 47.2, the **Purchaser** and the **Contractor** agree that no dispute shall be referred for resolution by any other means unless it has been referred to the **DRB** under Sub-clause 46.4 and:

(a) the **DRB** has failed to give its decision on the matter within the prescribed time; or

(b) a decision given within the prescribed time is either unacceptable to the **Purchaser** and/or the **Contractor** or has not been implemented within twenty-one days of the **DRB's** decision.

**46.6** Notwithstanding the existence of any dispute or any reference to the **Project Manager** under Sub-clause 46.3 or the **DRB** under Sub-clause 46.4, the **Purchaser** and the **Contractor** shall continue to perform their obligations under the **Contract**.

**46.7** The parties shall attempt in good faith to negotiate a settlement of any dispute or difference.

**46.8** If a dispute cannot be resolved by negotiation the parties may by agreement refer it to mediation in accordance with the procedures of the Centre for Effective Dispute Resolution (CEDR), the Chartered Institute of Arbitrators (CIArb) or some other appropriate body.

**46.9** No **Decision**, opinion, direction, or valuation given by the **Project Manager** shall disqualify him from being called as a witness and giving evidence before any third party, an adjudicator, **DRB**, **Expert** or arbitrator on any matter whatsoever relating to a dispute.

# 47. Adjudication

Delete the text of Sub-clause 47.2 in the General Conditions and insert:

**47.2** Notwithstanding any provision in the **Contract** for a dispute to be referred to a **DRB** in accordance with Sub-clause 46.4, to an **Expert** in accordance with Clause 48 (Reference to an Expert) or to arbitration in accordance with Clause 49 (Arbitration), either party shall subject to Sub-clause 47.1 of the **General Conditions** have the right to refer any dispute or difference (including any matter not referred to the **Project Manager** in accordance with Sub-clause 46.3) as to a matter under or in connection with the **Contract** to adjudication and either party may, at any time, issue a **Notice** (a '**Notice of Adjudication**') to the other stating his intention to do so. The ensuing adjudication shall be conducted in accordance with the edition of the 'Adjudication Rules' published by **IChemE** current at the time of service of the **Notice of Adjudication**.

# 50. Dispute Review Board

**50.1** Any dispute made referable to a **DRB** under Sub-clause 46.4 shall be determined by the **DRB** in accordance with the Rules for Dispute Review Boards published by **IChemE** current at the time of the referral.

**50.2** The **DRB** shall be appointed in accordance with the Rules for Dispute Review Boards.

**50.3** Any decision issued by the **DRB** shall not be binding on the parties. The **Purchaser** and the **Contractor** agree that any such decision shall not be admissible before any third party or arbitral tribunal on any matter whatsoever relating to the dispute.

*Alternative version of Sub-clause 50.3 if the decision of the DRB is to be binding but not final (see note above):*

**50.3** Any decision issued by the **DRB** shall be binding but not final, and the dispute shall be resolved by agreement or by reference to an **Expert** or to arbitration. The **Purchaser** and the **Contractor** agree that any such decision shall not be admissible before any third party or arbitral tribunal on any matter whatsoever relating to the dispute.

---

# Guide Note Q: Bonds and guarantees

Sometimes the Purchaser requires a guarantee or guarantees from a third party that the Contractor will perform the Contract and/or repay any money paid in advance if the Contract is terminated or the Contractor is in default. The third party may be one (or more) private persons or a bank, insurance company or business specialising in the giving of such guarantees. As guarantees are relatively expensive and often draw down on the Contractor's credit facilities, their requirement should be considered carefully for the particular Contract. These guarantees are generically referred to as bonds, but may take the form of insurance policies, bank guarantees, or repayment performance bonds. There may be confusion about these documents because different forms are required for different purposes; the basic purpose however is always to provide security to the Purchaser. Giving the bond or calling for payment under it does not affect the rights and obligations of the Purchaser or Contractor under the Contract or in respect of its breach or termination.

In general the procedures and forms used by the banks are followed to a greater or lesser extent by the other bodies which provide such guarantees.

Banks normally only issue 'on-demand' guarantees while insurance or surety companies normally only issue conditional guarantees. In all cases the bondsman's only interest is that the requirements of the bond are met. In the case of an on-demand bond, the bondsman will not enquire into or adjudicate upon whether the Contract has been correctly performed. The bondsman's duty is merely to see that the demand is in correct form as specified in the bond. On receipt of such a demand the bondsman is bound to pay, unless it has notice of provable fraud, and will certainly have an indemnity whereby it can recover the money from the Contractor. In contrast, a conditional bond may require evidence (for example, a Court judgement or an arbitrator's award) of default by the Contractor.

In this Guide Note the words 'bondsman' and 'bond' refer to all givers of guarantees and the guarantees given respectively.

Typical terms can include:

(a) The maximum amount payable being expressed in both words and figures;
(b) A calendar date for expiry of the bond (a period such as three hundred and sixty five days after Taking Over will not be acceptable because the bondsman cannot know for sure the starting date of the three hundred and sixty five days). To allow for time to call the bond the expiry date should be after the latest date the obligation guaranteed is likely to expire (e.g. thirty days);
(c) A statement that the bond is irrevocable;
(d) A 'waiver' clause that the bond is not affected by changes in the Contract, extensions of time or other indulgence;
(e) A requirement to return the bond for cancellation when it has expired or is no longer required under the Contract;
(f) A law and exclusive jurisdiction clause.

It follows that the wording of the operative part of the bond must describe exactly the document(s) to be presented as, or with, the demand for payment so as to reflect precisely the purpose of the bond and the intentions of the parties. If the bond is to be given during the effective period of the Contract, a Special Condition will be needed stating when the bond is to be given and how the wording is to be settled, for example, by including a pro-forma bond in an annex to the conditions of contract. It is normal for the Special Condition to state what the consequence will be if the bond is not furnished.

An appropriate Special Condition requiring the provision of a bond might read as follows. The form annexed to the Conditions of Contract would show whether it was an on-demand or conditional bond.

---

**EXAMPLE CLAUSE**

# XX. Provision of performance bond

**XX.1** Prior to the commencement of the **Works** the **Contractor** shal at his own expense furnish to the **Purchaser** an irrevocable guarantee/bond from a surety who is a first class bank or insurance company or other surety or sureties acceptable to the **Purchaser** (such acceptance shall not be unreasonably withheld) in an amount of *[currency and amount in words and figures]* and in the form annexed to these **Contract Conditions**. Failure to provide such a bond shall be a material breach of the **Contract** within the meaning of Sub-clause 44.2 (c) and Clause 44 (Termination for Contractor's default) shall apply.

---

The purposes for which bonds are required vary considerably but the following are common.

**Tender bonds**
Tender bonds, also referred to as bid bonds, are issued at the tendering stage. The purpose is to ensure that the successful tenderer enters into a Contract or pays the Purchaser his expenses of abortive negotiations. It may be payable on a written demand stating that the tenderer was informed that his tender was acceptable but nevertheless within a stated period failed, neglected or refused to enter into a binding Contract.

**Down payment or advance payment bonds**

The purpose is to secure repayment of the down or advance payment (or a suitable part thereof) if the Contract is terminated, frustrated or otherwise comes to a premature end. Payment may be on written demand of the Purchaser without proof or conditions but the amount payable should reduce in steps by stated amounts or percentages as the Contract proceeds, normally in proportion to instalments of the Contract Price payable and at the dates the instalments are due. A Special Condition should specify the procedure for notification to the bondsman by either party of the payments due and any requested supporting documents to evidence the reduction in value of the bond.

**Performance bonds**

The purpose is to secure proper performance of the Contract by the Contractor and to provide the Purchaser with an alternative source of funds to cover any extra cost of having the Contract performed by another contractor and/or to meet the Purchaser's entitlements for financial loss including any liquidated damages if the Contractor fails to perform. The amount is typically five to ten per cent of the Contract Price. A performance bond can be either on-demand or conditional.

**Retention bonds**

*See Schedule 19 (Contract Price and payment)*

Purchasers would usually prefer to make the final payment or impose retention on all payments earned until after the results of performance tests are known and after the Contractor is finally released from his liabilities for Defects. It is recognised that this adversely affects the Contractor's cash flow and prevents him from recovering all his costs and realising his profit when all, or virtually all, of his work has been done. It is therefore common practice to pay the final instalment or release the retention amount on taking over or completion of commissioning provided that the Contractor furnishes an 'on demand' bond for an amount equal to the final payment or retention amount with an expiry date no earlier than the date on which the Contractor is released from his liabilities for Defects.

**Parent company guarantees**

Parent company guarantees may be required when a party to the Contract is a subsidiary or company specially set up for the Contract (for example, a company or a company to trade in a particular country or a jointly owned consortium company) with insufficient financial assets to meet the full extent of his financial obligations. This usually applies to the Contractor (and its parent company) but may also apply to the Purchaser where his only assets are the completed Plant and a guarantee is necessary to support the Purchaser's payment obligations.

## Guide Note R: Rights of third parties

Under English law a person who is not a party to a contract ('a third party') can enforce his rights only to the extent that the Contracts (Rights of Third Parties) Act 1999 permits.

In the context of a contract between a Purchaser and a Contractor, a Subcontractor is a third party. In the same way, in relation to a subcontract, the Purchaser is a third party. Either Purchaser or Subcontractor may be able to show that the contract (which includes a subcontract for this purpose) gives him certain rights which he may wish to exercise directly against one of the parties to the contract.

Sub-clause 9.6 expressly permits any Subcontractor to exercise rights directly against the Purchaser in respect of any misuse of Confidential Information (see also Sub-Clause 20.3) and in respect of the indemnities given in Sub-clauses 30.11 and 30.14. Similarly Sub-clause 9.7 is intended to open the way by requiring the Contractor to draft his Subcontracts appropriately for the Purchaser as a third party to those Subcontracts to enforce directly the Purchaser's rights to the Subcontractor's obligations concerning Confidential Information and Defect rectification. Such provisions will be upheld by the Contracts (Rights of Third Parties) Act 1999. See also Sub-clause 9.8.

These rights would be in addition to any rights of action between the Purchaser and Contractor.

The advantage is that the parties most directly affected, the Purchaser and the Subcontractor, can deal directly where there has been a breach of a confidentiality undertaking, or where an item of Plant supplied by a Subcontractor develops a Defect.

Some Purchasers prefer not to rely on the Contracts (Rights of Third Parties) Act 1999 and instead enter into a direct agreement with a Subcontractor. Such an agreement is usually called a Collateral Warranty or Collateral Agreement. It is an agreement in addition to the chain of main contract and subcontract, which allows the Purchaser to enforce the Subcontractor's obligations relating to Defect rectification and Confidential Information and possibly other obligations directly as an alternative to going through the Contractor. It allows the Subcontractor to enforce directly those of the obligations of the Purchaser which are expressed to be for the Subcontractor's benefit, for example, those relating to Confidential Information and certain indemnities.

Under projects that are funded by banks or that are likely to be sold on to other parties, it is usual for the Purchaser to require that the Contractor enter into Collateral Warranties/Agreements for the benefit of such funders.

Appropriate legal advice will be necessary as to the drafting of such a Collateral Warranty/Agreement, together with an appropriate Special Condition which will impose an obligation on the Contractor to ensure that his Subcontractors are bound to enter into such an agreement with the Purchaser, or that the Contractor will enter into such an agreement with funders.

## Guide Note S: Joint ventures

It is often the case on large projects that the role of Contractor is undertaken by more than one contractor, working together in a joint venture. A joint venture can be formed in two ways:

- Where two or more contractors form a new company or other corporate body, or
- Where two or more contractors trade together under a contractual joint venture agreement which does not provide for the formation of a separate company.

The first approach still provides for a single company to contract with the Purchaser, albeit the Purchaser will probably wish to have guarantees from each of the shareholder contracting companies that have together formed the new company.

However, in terms of the use of the General Conditions the new company is the Contractor.

The second approach is somewhat different, as now two or more companies are going to fulfil the role of Contractor and all or only one will be directly in contract with the Purchaser. The way in which such contractual arrangements are structured can be complex and will require specialist advice as to how the Agreement is to reflect the requirements of the several parties to the Contract. Usually the Purchaser will require each member of the joint venture to be jointly and severally liable for the performance of the Contractor's obligations under the Contract. The Purchaser may also require details of the joint ventures management structure and how they will carry out the work between them.

# Appendix A: Performance Test Flowcharts

These flowcharts are intended to assist in the understanding of Clause 35 (Performance Tests). Readers should refer to that clause for a definitive statement of the parties' legal obligations.

## Preparation for Tests

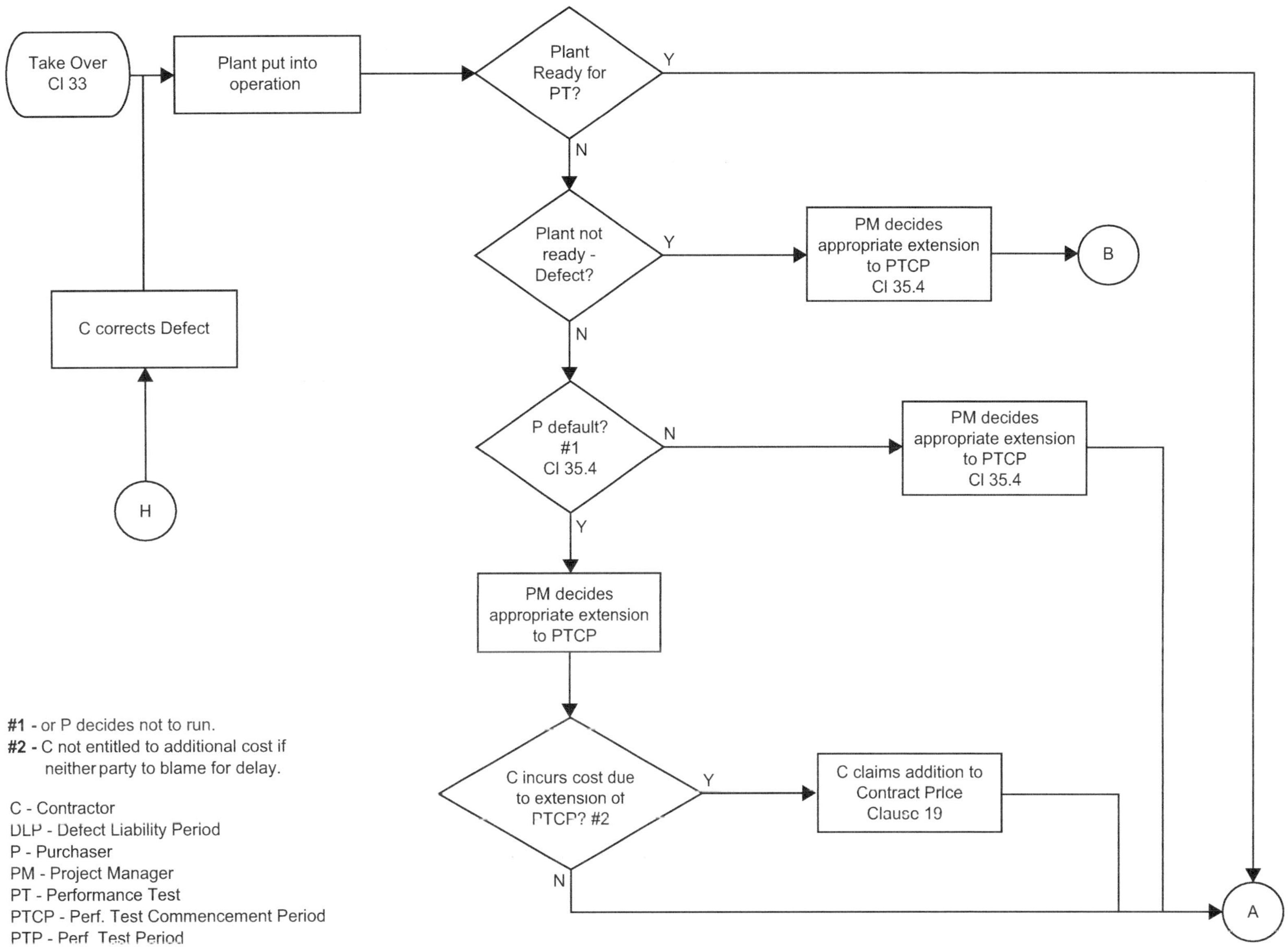


**#1** - or P decides not to run.
**#2** - C not entitled to additional cost if neither party to blame for delay.

C - Contractor
DLP - Defect Liability Period
P - Purchaser
PM - Project Manager
PT - Performance Test
PTCP - Perf. Test Commencement Period
PTP - Perf. Test Period

## Test Commencement

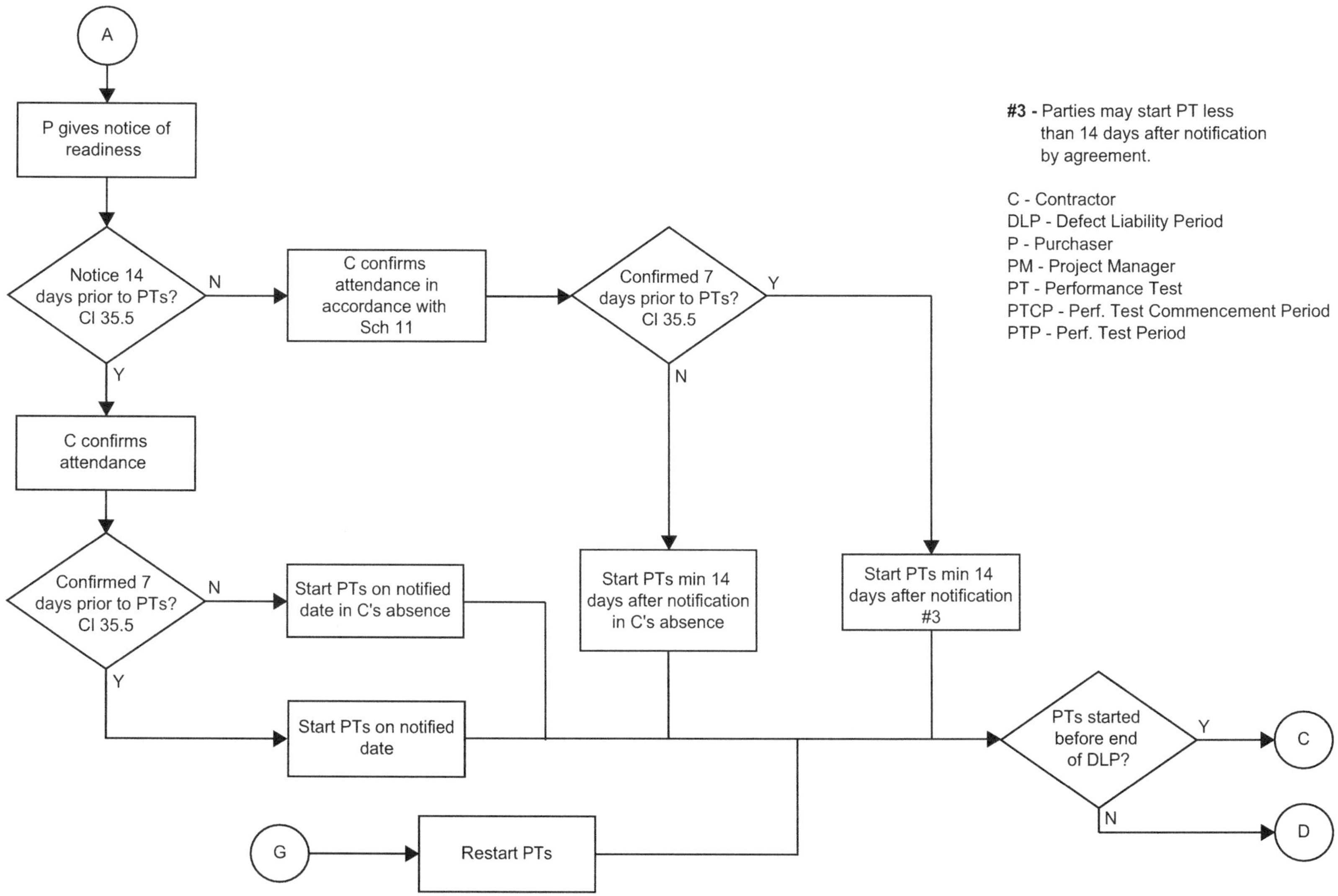


**#3 -** Parties may start PT less than 14 days after notification by agreement.

C - Contractor
DLP - Defect Liability Period
P - Purchaser
PM - Project Manager
PT - Performance Test
PTCP - Perf. Test Commencement Period
PTP - Perf. Test Period

## Completion and Evaluation

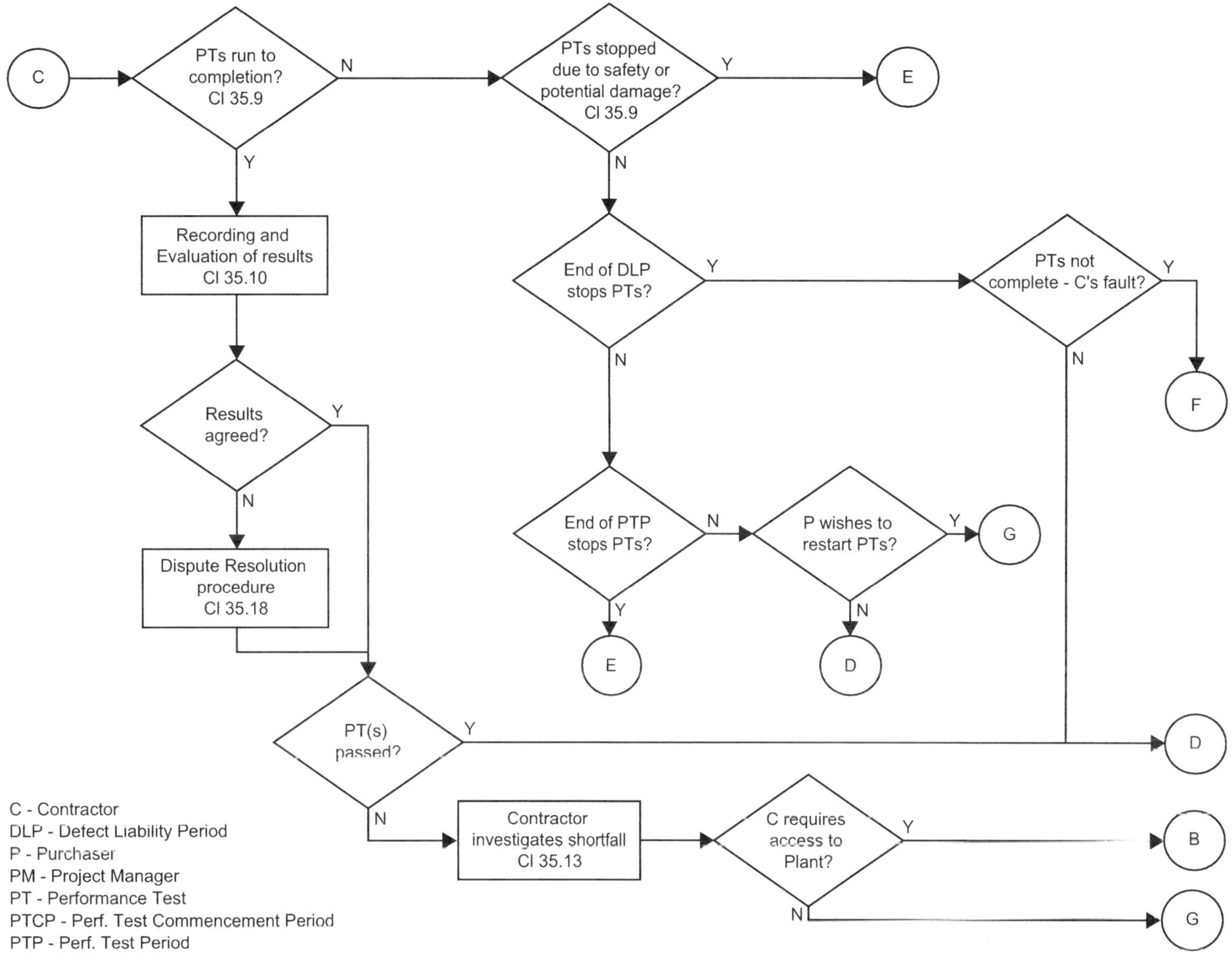


C - Contractor
DLP - Defect Liability Period
P - Purchaser
PM - Project Manager
PT - Performance Test
PTCP - Perf. Test Commencement Period
PTP - Perf. Test Period

## Corrections to Plant and Extension to Test Period/1

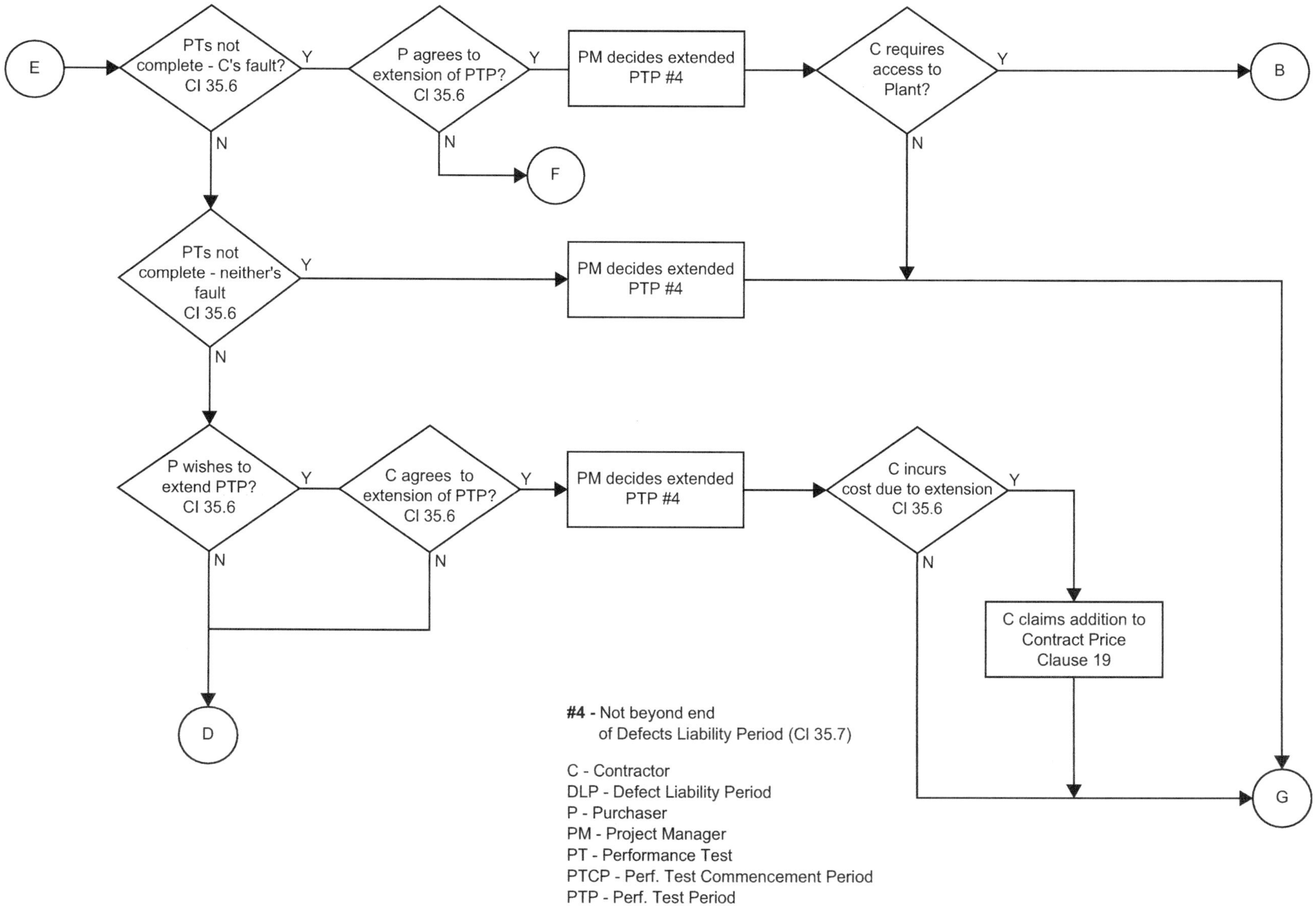

## Corrections to Plant and Extension to Test Period/2

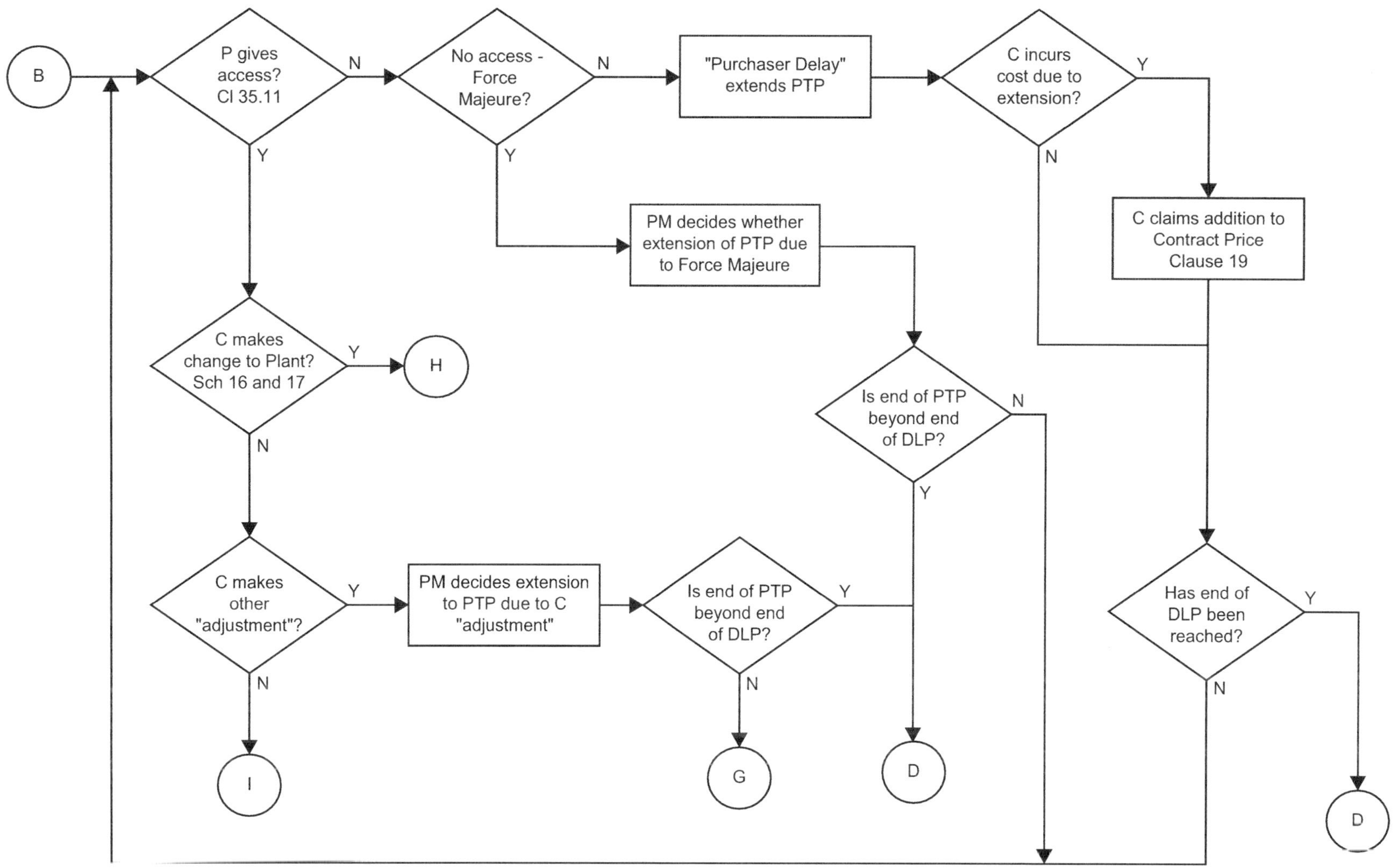


C - Contractor
DLP - Defect Liability Period
P - Purchaser
PM - Project Manager
PT - Performance Test
PTCP - Perf. Test Commencement Period
PTP - Perf. Test Period

## Damages and Acceptance

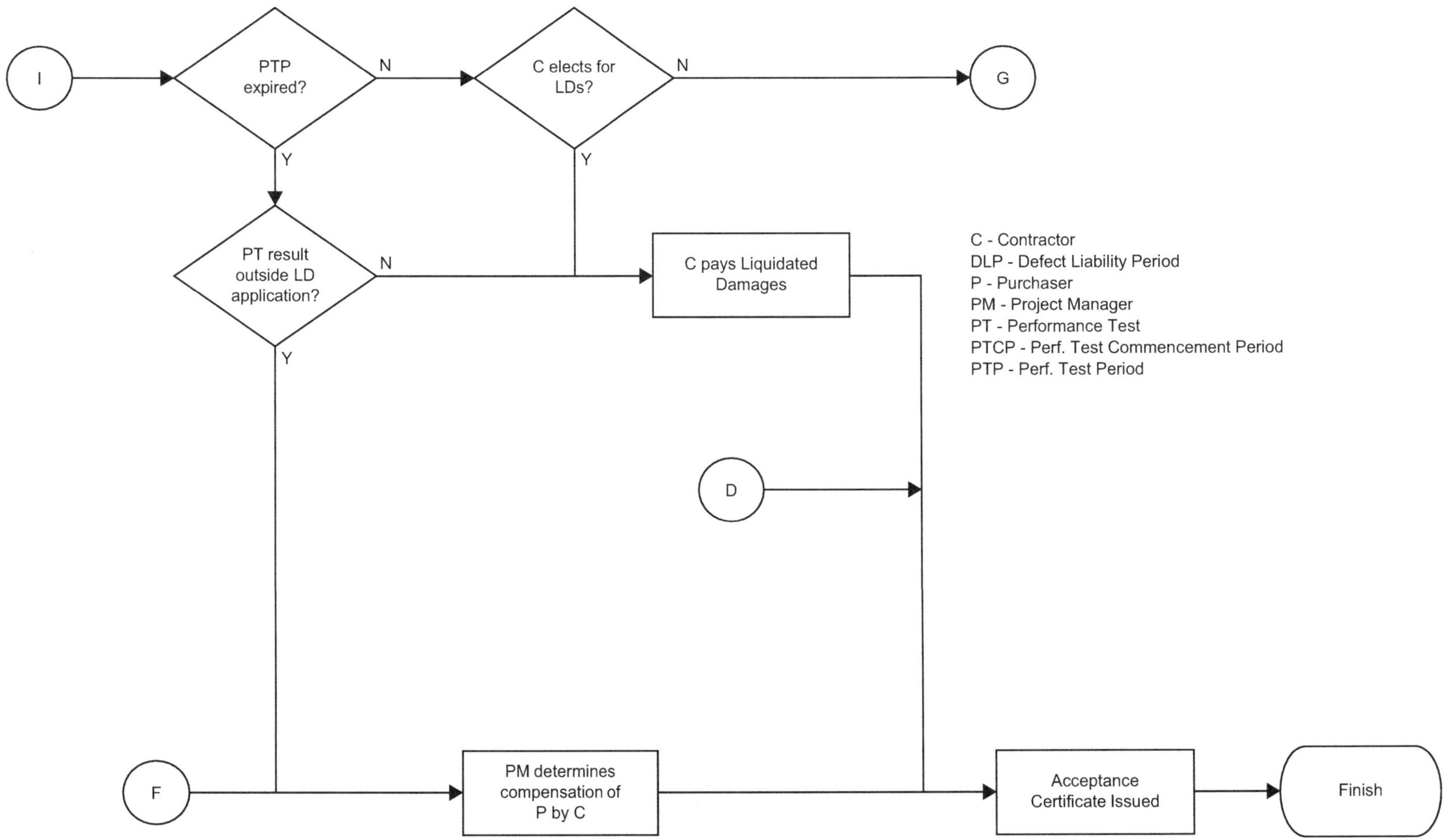

# Appendix B: Sample Certificates

The following sample certificates are given as an indication only of the likely content of each certificate. Users should take care to ensure that the words used in a certificate accurately reflect the terms of the Contract as they apply to the particular situation in which a certificate is issued.

## Sample Construction Completion Certificate

Contract Title and Identification Number:

Purchaser:

Contractor:

List of parts of the Plant substantially completed, covered by this Certificate:

Criteria listed in Schedule 14 which apply to these parts of the Plant:

Programme for inspection and tests:

In my opinion the parts of the Plant listed above are substantially complete and meet the above-mentioned criteria listed in Schedule 14. In accordance with Sub-clause 32.2 of the General Conditions I therefore propose the above programme for inspection and tests commencing on *[date]*.

Signed: ....................................

Name: ....................................

Date: ......................................

On behalf of the Contractor

**Project Manager's certification:**

* In accordance with Sub-clause 32.7 of the General Conditions the parts of the Plant listed above do not meet the criteria in Schedule 14 as listed above in the following significant respects:

* In accordance with Sub-clause 32.6 of the General Conditions I hereby certify that the parts of the Plant listed above were substantially complete in accordance with Sub-clause 32.6 of the General Conditions as of *[date]* and in such a condition that any procedures needing to be carried out before the Plant is put into operation may be safely carried out.

* In accordance with Sub-clause 32.6 of the General Conditions I require the following minor items to be completed before the issue of a Take Over Certificate covering the parts of the Plant listed above:

Signed: ....................................

Name: ....................................

Date: ......................................

Project Manager

**Delete as applicable*

# Sample Take Over Certificate

<u>Contract Title and Identification Number:</u>

<u>Purchaser:</u>

<u>Contractor:</u>

<u>Date of Certificate</u>

**(Alternative A—where the Plant or any Section is being Taken Over under Sub-clause 33.7 or 33.10)**

(1) In accordance with Sub-clause *[33.7]/[33.10]* of the General Conditions I hereby certify that the *[Plant]/[Section(s) of the Plant listed below]*, apart from any parts of the *[Plant]/[Section(s) listed below]*, *[has]/[have]* satisfied the requirements of the Specification and Schedule 15 of the Contract and the Contractor has complied with Sub-clause 21.9 of the General Conditions and therefore such *[Plant]/[Section(s) of the Plant listed below] [is]/[are]* Taken Over by the Purchaser as from the Taking Over time and date stated below.

**Sections of the Plant Taken Over:**

(2) The following parts of the *[Plant]/[Sections]* are excluded from Taking Over

(3) The following minor items of work are still to be completed in accordance with Sub-clause 33.8 of the General Conditions,

(a) .....................

(4) The following Take Over procedures have been omitted by operation of Sub-clause 33.10 of the General Conditions and still remain to be carried out by the Contractor in accordance with Clause 33 (Taking Over) of the General Conditions.

(a) .....................

(b) .....................

(5) The Taking Over time and date is:

**(Alternative B—where the Plant or any Section is being Taken Over under Sub-clause 33.9)**

(1) In accordance with Sub-clause 33.9 of the General Conditions I hereby certify with the Contractor's consent that the *[Plant]/[Section(s) of the Plant listed below]* apart from any parts of the *[Plant]/[Section(s) listed below] [is]/[are]* Taken Over by the Purchaser as from the Taking Over time and date stated below.

**Sections of the Plant Taken Over:**

(2) The following parts of the *[Plant]/[Section] are* excluded from Taking Over for the reason(s) stated:

(a) because the take over procedures have not been carried out: ......

(b) because the take over tests have not been passed: ......

(3) The following minor items of work are still to be completed in accordance with Sub-clause 33.8 of the General Conditions.

(a) .....................

(b) .....................

(4) The following take over procedures have been omitted by operation of Sub-clause 33.9 of the General Conditions and still remain to be carried out by the Contractor in accordance with Clause 33 (Taking Over) of the General Conditions.

(a) .....................

(b) .....................

(5) The Taking Over time and date

Signed: ................................

Name: ................................

Project Manager

# Sample Acceptance Certificate

Contract Title and Identification Number:

Purchaser:

Contractor:

Date of Certificate:

(1) In accordance with Clause 36 of the General Conditions I hereby certify that the *[Plant]/ [Section(s) of the Plant listed below]* have:

*(a) passed all the performance tests satisfactorily;

*(b) passed all the performance tests, except for those performance tests for which the appropriate *[liquidated damages have/compensation has]* been paid or allowed for by Contractor

and I therefore issue this Certificate in accordance with Sub-clause 36.3 of the General Conditions;

*(c) been Accepted by the Purchaser by a notice of election issued to the Contractor in accordance with Sub-clause 36.5 of the General Conditions.

*(2) Known Defects to be made good by the Contractor and minor items still remaining to be completed in accordance with Sub-clause 36.3 of the General Conditions are listed in the attached 'punch list'.

Signed: ...................................

Name: .....................................

Project Manager

**Note: delete as applicable.*

## Sample Final Certificate

Contract Title and Identification Number:

Purchaser:

Contractor:

Date of Certificate:

(1) In accordance with Sub-clause 38.1 of the General Conditions I hereby certify that in respect of the *[Plant]/[Section(s) of the Plant listed below]*:

(a) the Defects Liability Period has expired;

(b) the Contractor has made good all Defects that have appeared during the Defects Liability Period ; and

(c) the Acceptance Certificate has been issued in respect of the *[Plant]/[Section(s) of the Plant listed above].*

(2) This Final Certificate is effective from *[time and date]*

*(3) The following items are excluded from this Final Certificate in accordance with Sub-clause 38.2 of the General Conditions.

(a) ................................

(b) ................................

*(4) This is the last Final Certificate to be issued under the Contract.

Signed: ................................

Name: ..................................

Project Manager

**If this is the last Final Certificate, (3) should be deleted. If this is not the last Final Certificate, (4) should be deleted.*

# Index

Abbreviations used under the heading 'other' are as follows: Agt. = Agreement; App. = Appendix; GN = Guide Note; IN = Introductory Note; Sch. = Schedule; Specn. = Specification.

## Q



## R